Developments in Animal and Veterinary Sciences, 2

Ethology of free-ranging domestic animals

Developments in Animal and Veterinary Sciences, 2

Ethology of free-ranging domestic animals

G.W. Arnold

CSIRO Division of Land Resources Management, Western Australian Laboratories, Wembley, W.A., Australia

and

M.L. Dudzinski

CSIRO Division of Mathematics and Statistics, Canberra, A.C.T., Australia

ELSEVIER SCIENTIFIC PUBLISHING COMPANY
Amsterdam — Oxford — New York 1978

ELSEVIER SCIENTIFIC PUBLISHING COMPANY
335 Jan van Galenstraat
P.O. Box 211, 1000 AE Amsterdam, The Netherlands

Distributors for the United States and Canada:

ELSEVIER/NORTH-HOLLAND INC.
52, Vanderbilt Avenue
New York, N.Y. 10017

Library of Congress Cataloging in Publication Data

Arnold, G W
Ethology of free-ranging domestic animals.

(Developments in animal and veterinary sciences ; 2)
Bibliography: p.
Includes index.
1. Domestic animals--Behavior. 2. Ruminants--Behavior. 3. Grazing. 4. Range management.
I. Dudzinski, M. L., joint author. II. Title.
III. Series.
SF756.7.A76 636.08'3 78-13776
ISBN 0-444-41700-1

ISBN 0-444-41700-1 (Vol. 2)
ISBN 0-444-41703-6 (Series)

Printed in The Netherlands

CONTENTS

PREFACE

Man has domesticated relatively few grazing animals and he has selectively bred them to suit many climatic and husbandry situations. In many parts of Europe and North America both cattle and sheep, but particularly cattle, spend only part of their lives grazing at large, and even then it is often under restricted conditions. In other parts of Europe, the Americas, Africa and Asia, herds and flocks of camels, goats, sheep, cattle, llamas and alpacas are herded and shepherded. In contrast to this, husbandry systems which allow year-round grazing, with complete freedom of movement, predominate on the grasslands and rangelands of Australia and in some other parts of the world.

This book is concerned with the patterns of behaviour of such free-grazing domestic animals; how these patterns are modified by environmental and husbandry practices; and the effects that interactions between behaviour and husbandry have on the environment.

We give no background in the fundamentals of behaviour as these are written of elsewhere. However, we believe that a reader without such a grounding in the fundamentals will be able to follow our ideas. Our aim is not to wrap up descriptions of patterns of behaviour in ethological jargon and theory, indeed the book may disappoint the behavioural scientist because it does not. It is written in the hope that animal scientists, ecologists, and students may gain a better insight into the behaviour of free-grazing domestic animals. Since in our studies of behaviour we have tried to be quantitative wherever possible, we attempt, in this book, to critically assess the available information on each topic on the basis of ethological soundness and quantitative rigour.

We are gratefully indebted to our colleagues Drs Elizabeth Shillito-Walser, K.R. Croker, W.A. Low, V.R. Squires and especially Mrs Doris Leadbetter for their criticisms of the manuscript. Photographs in the book are taken by the authors and by Messrs F. Dandridge and C. Totterdell. Our especial thanks are due to Mr K. Short for preparing the many illustrations. We acknowledge the courtesy of the authors and publishers in allowing the reproduction of numerous figures.

INTRODUCTION

Man domesticated animals for specific purposes and has, over thousands of years, selectively bred them to fit more closely his requirements. In the course of time, man's husbandry practices have changed and so he continues to adapt his animals to changing conditions. New areas of the world have been developed, and still are being developed, from virgin lands. Existing species of domestic grazing animals — sheep, goats, cattle, horses, buffalo, reindeer, camels, llamas and alpacas — are introduced as exotic species into these new environments.

Two approaches have been used in seeking adaptation — selection or breeding of suitable strains (breeds), and using a strain which has a great plasticity and can adapt to many environments. One example of the latter approach is the Merino sheep in Australia. In the course of 175 years since its introduction from Spain, it has been successfully used for wool production in dry tropical, arid, dry temperate, sub-humid temperate, humid temperate and sub-alpine climatic zones. The types of vegetation it flourishes or survives on vary from temperate man-made grasslands to natural grasslands, spinifex, and saltbush shrublands. Only for the tropical areas is it necessary to develop specific strains. The Friesian breed of cattle is another example of wide adaptation.

In other cases many breeds have been developed within the one species, as with sheep in Britain, where the claim of the protagonists for a particular breed is that it is the one best adapted to their particular area. Such claims, not often scientifically established, would seem to imply a need to develop a breed for a particular ecological niche. It seems, though, that it is only the extremes of heat, humidity and cold that demand more careful breeding and selection of breeds of a species.

The great range of conditions to which some breeds can adapt depends in large part on the adaptation of behaviour of the animals to different husbandry, nutritional, climatic and other environmental conditions. Husbandry practices may vary from close herding during the day and enclosure or kraaling at night, to intensive housing, or to free ranging under a wide variety of grazing management practices. In this book, we are concerned only with behaviour under free-ranging conditions and the ways in which the behaviour of the various species differs. This requires a description of the basic pattern of feeding and other daily maintenance behaviour, social behaviour, mating and maternal behaviour and how these patterns are varied under different environmental and husbandry conditions. The limits to adaptation of behaviour need to be considered together with reasons for the limits. For example, it may be possible for the diurnal pattern of grazing and resting to be changed to reduce physiological stress due to heat, without affecting the nutrition of the animal. In addition, the animal may graze in areas which provide a considerable degree of shelter from cold, or it may seek exposed windy areas that reduce its heat load. It is important to know whether such adaptation in behaviour occurs and under what conditions. It is certain that, beyond some limits, the need to minimise physiological stress will be greater than the nutritional needs and the animal will reduce the time it spends feeding.

All animal communities have a social structure. Domestic herds and flocks have to develop structures that are quite different from those of their wild ancestors because man often keeps separate both ages and sexes of a species. Also, he often reconstitutes his herds and flocks, thus bringing strangers together and separating individuals that have close bonds. What are the social structures in free-ranging herds and flocks? What social disruption occurs when strangers are introduced or bonds are broken? Such questions need to be answered in terms of the behaviour of the animals and, importantly, their productivity for man.

Reproduction is usually closely controlled by man in his domestic animals by controlling the time males and females are kept in contact and by altering the ratio of males to females. Thus it is necessary to consider how mating behaviour is influenced by these controls and by other conditions such as the space — area, topography, supplies of food and water — available to the animals. Success in rearing their young depends in large part on the behaviour of the dam at parturition and in the critical post-parturition period. Under free-ranging conditions, shelter and isolation-seeking behaviour could be important in determining survival as well as reaction to predators.

The most basic requirements of food and water are under the control of man as well as the environment, but under free-ranging conditions the ability of the animal to select an adequate diet and the distance it is able or willing to travel to water will determine both survival and productivity. Thus the feeding behaviour and the way in which animals use the environment are given considerable attention.

The behaviour interactions between the animal and its environment, including man, form the central theme of this book. They are important because they should provide guides to better husbandry and productivity.

Each pattern of behaviour has a definite special adaptive function, which can generally be related to nine general functions as stated by Scott (1969): ingestive; eliminative; sexual; care-giving (epimeletic); care-soliciting (et-epimeletic); agonistic; allelomimetic; shelter-seeking; investigatory and behaviour disorders (termed by some authors anomalous or maladaptive behaviour). These patterns of behaviour have been grouped in this book so that ingestive and eliminative behaviour and shelter-seeking behaviour are considered as basic daily activities necessary for the maintenance and survival of the animal. They are covered in Chapters 1 and 3. Agonistic and allelomimetic and investigatory behaviour are mainly concerned with social organisation and the use of the environment and are largely dealt with in Chapter 2. Sexual behaviour is considered in Chapter 4 and maternal behaviour, which includes care-giving and care soliciting behaviour, is the subject of Chapter 5. Behavioural disorders are dealt with in relation to maternal behaviour, but appear infrequently in other types of behaviour in free-grazing animals. The final chapter, Chapter 6, considers the interactions of behaviour and husbandry on animal productivity and the environment. It also gives attention to behavioural characteristics desirable in domestic animals in that they might be used in management of amenity areas that are of increasing importance in the more affluent parts of the world.

Some of the aspects presented have been considered by various authors more briefly in "The Behaviour of Domestic Animals" (Hafez, 1969). Squires (1975a) has given a brief review of the behaviour of domestic sheep, and Fraser (1968) has considered reproductive behaviour in ungulates in detail. Most of these previous publications are concerned with both restricted and free-grazing animals, and detailed ethograms are given for some species of such specific activities as rumination, courtship and mating, and agonistic behaviour. We have not repeated them here because they are common to the animal whether it is free or restricted.

Unfortunately, information on aspects of behaviour varies in both quantity and quality. Some species, like the goat, camel, reindeer, llama and alpaca have been almost totally ignored by ethologists. Even with sheep, cattle and horses there is a marked difference in the information available on specific types of behaviour. For example, maternal behaviour has been studied intensively in sheep but only rarely in cattle and horses. Eliminative behaviour is well documented for cattle, but not for sheep. Where possible, comparisons are made between the behaviour of different species, and between domesticated and wild or feral herds and flocks.

Chapter 1

DAILY MAINTENANCE BEHAVIOUR

The activities which each and every animal does every day are those concerned with maintenance and survival — grazing, browsing and/or feeding on supplements, walking, ruminating, resting, defaecating and urinating. There are distinct diurnal patterns for some activities but not for others. However, all are linked into a daily pattern of changing activities.

In part, the diurnal pattern will depend on the digestive physiology of the species. Those species that ruminate (cattle, sheep, goat, camels) need time for rumination, whereas horses, in which digestion of forage takes place in the caecum, do not. It will also be influenced by weather conditions and the need to minimise physiological stress due to heat or cold and by the internal demands of the individual animal. Some activities, like feeding, will occur daily throughout the year; others, like drinking and sheltering from heat or cold, may occur daily or only at certain times of the year. An example of the daily activities of a cow in summer is given in Fig. 1.1.

In this chapter the diurnal patterns of these activities, the duration or frequency of them and the effects of various factors on the patterns, time or frequency are considered.

GRAZING

Diurnal pattern

The effects of daylength

Both sheep and cattle are more active during the daylight than during darkness where grazing is concerned. Most studies have shown that a major grazing period begins around sunrise and another before sunset: in the hours between these periods, and at night, other and shorter periods of grazing take place. The timing of these 'secondary' periods is influenced mainly by several climatic factors and by grazing pressure.

Fig. 1.2 shows the diurnal variation in the intensity of grazing over the year for sheep at Canberra, Australia. Similar results were obtained in England for sheep and cattle (Hughes and Reid, 1952). There are some periods of the day when all the animals in a flock or herd will be grazing and others when only part of the flock will be grazing. Generally, in latitudes greater than 35° the breaks between grazing decrease as the days get shorter, until in

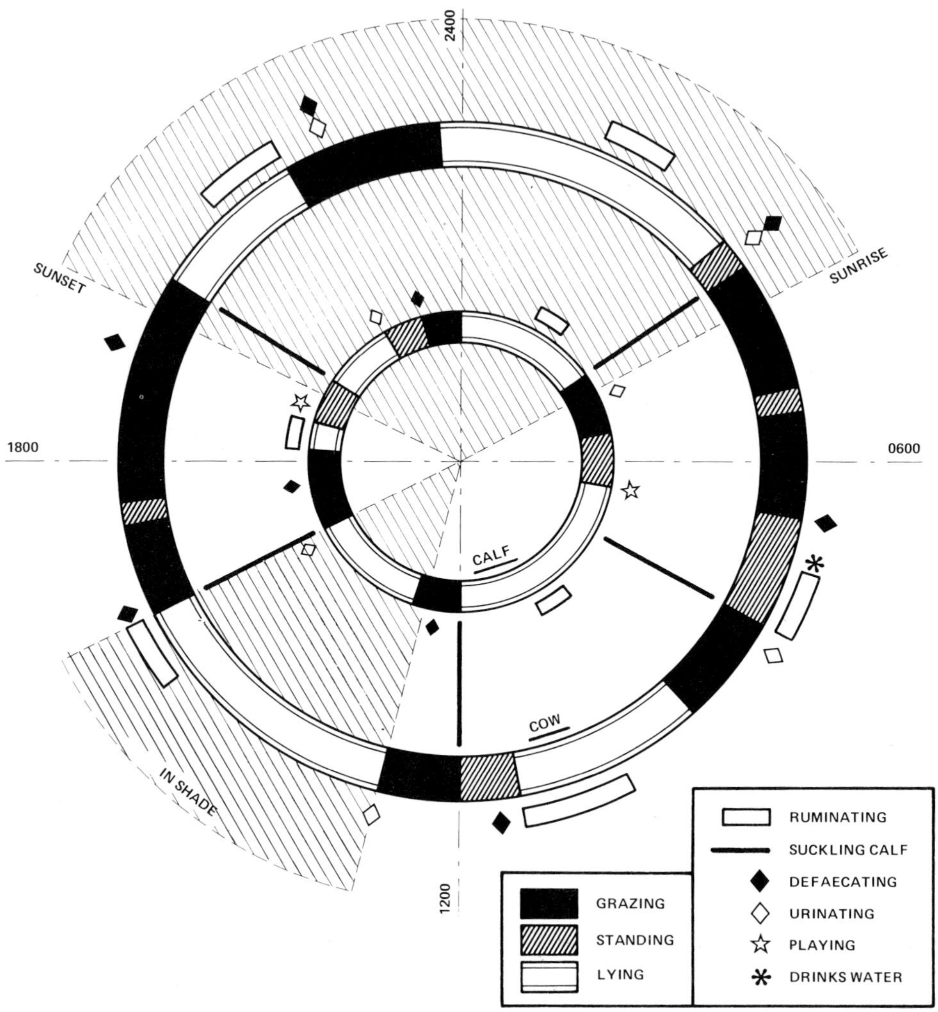

Fig. 1.1. Ethogram of the daily activities of a cow and calf.

mid-winter some animals will always be found grazing during daylight.

Horses show much less diurnal variation in grazing time. Comparisons between sheep, cattle and horses (Arnold, unpublished) are shown in Fig. 1.3. Each plot is a mean of four (summer) and six (winter) observations. Sheep and cattle had very similar patterns but the horses in both winter and summer did much more grazing at night than the sheep and cattle.

In tropical climates, where daylength varies little, some studies have shown that cattle graze predominantly at night (Payne et al., 1951); others

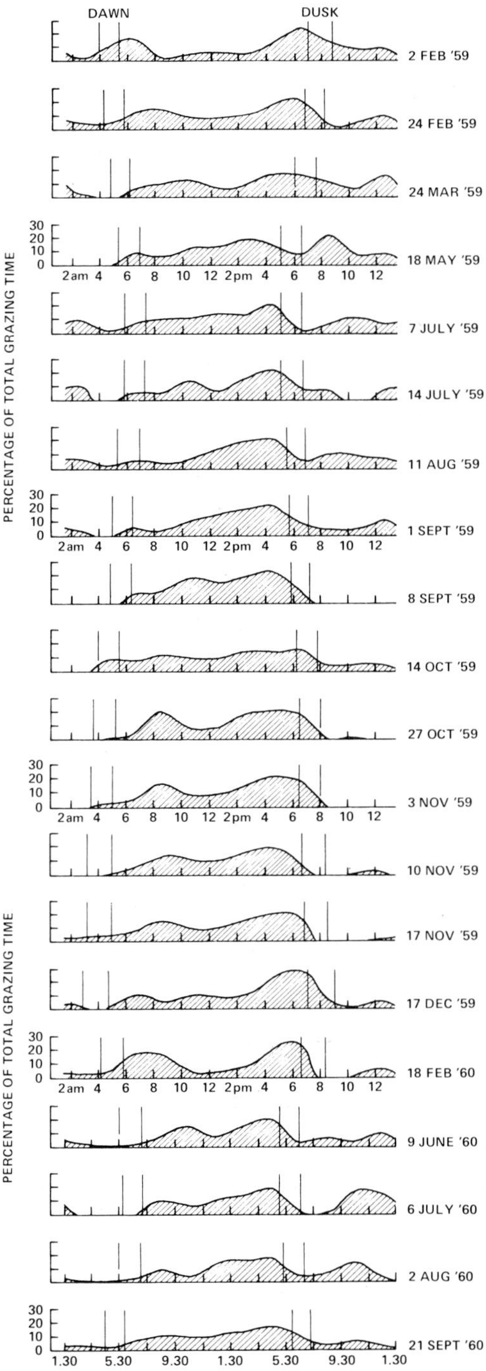

Fig. 1.2. Seasonal changes in the diurnal pattern of grazing of sheep at Canberra, Australia. (Source: Arnold, 1962.)

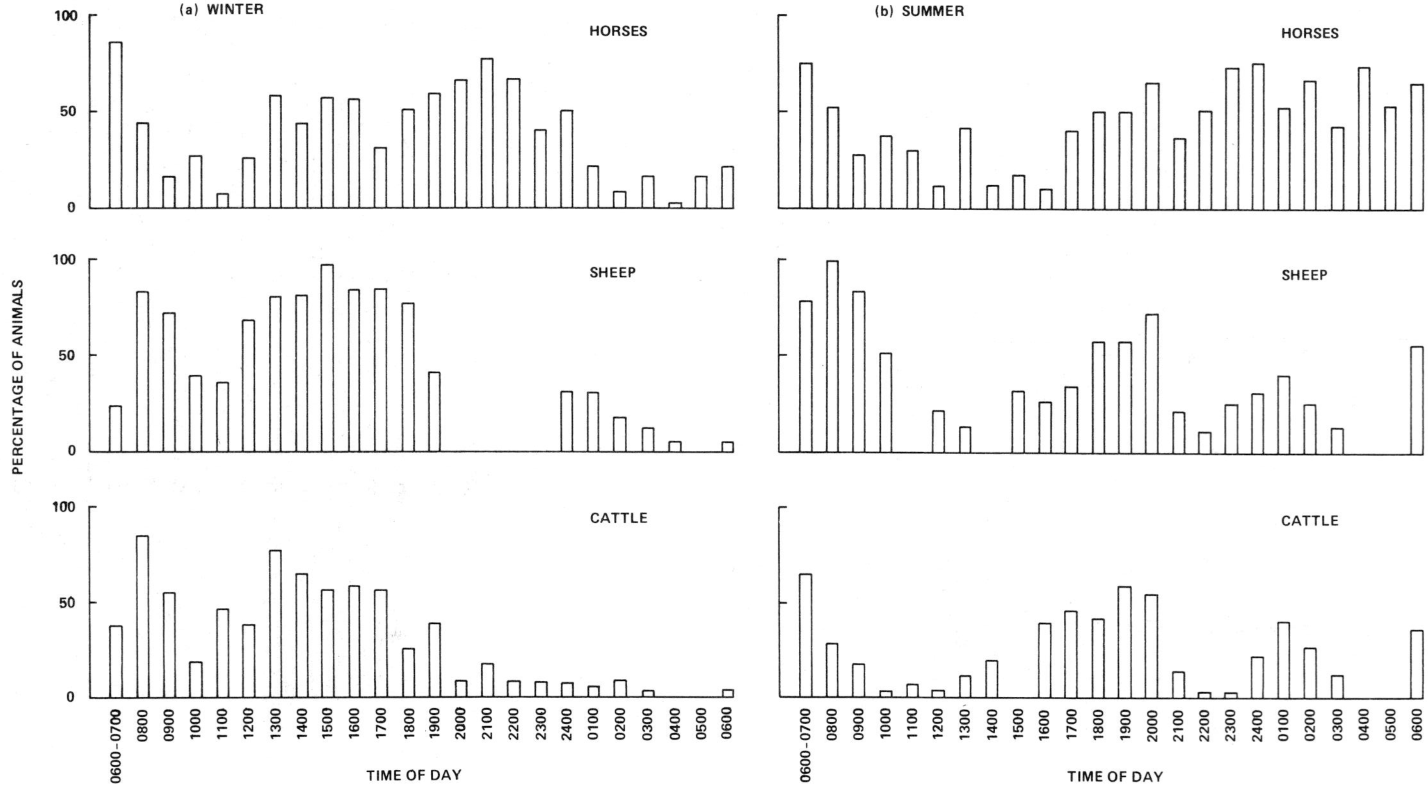

Fig. 1.3. Diurnal pattern of grazing of horses, sheep and cattle in (a) winter and (b) summer at Baker's Hill, Western Australia.

show this is not always the case (Harker et al., 1954, 1961). It may be that the stresses of high temperature and high humidity during the day cause different discomfort to cattle of different breeds.

Data from all published studies we found on grazing behaviour in cattle have been collated, and the relationships between the proportion of grazing done at night and daylength, latitude, and total grazing time examined: in all, 131 sets of data were obtained.

Regression analysis showed that the proportion of night grazing was significantly related to total grazing time (+ve), daylength (—ve) and latitude (—ve), with daylength being the most important. When data for the tropics were omitted, the significance of latitude disappeared, and so it seems that animals are influenced by other climatic factors in the tropics. It must be pointed out, however, that the regressions accounted for only 45% of the variation in the proportion of grazing time occurring at night.

The effects of temperature and humidity

Unfortunately, in very few studies have both temperature and humidity been recorded. Analysis of data for 25 studies where temperatures were recorded showed that the greater the differences between maximum and minimum temperature, the less was night grazing. For the 16 studies where humidity data were also available, fluctuation in humidity was important, there being more night grazing when humidity in the day was high. A specific example is quoted by Goldson (1976). He says that Lampkin et al. (1958) found that steers grazed at Muguga on the equator, where the climate is temperate due to the high altitude, did only 8% of their grazing at night but when moved to hot humid coastal environment did 29% of their grazing at night.

The time when animals graze during the day is influenced by weather conditions. Dwyer (1961) observed that cattle grazing rangeland moved as they grazed in the morning from one watering point to another and the hotter it was, the earlier they arrived at the second watering point; they appeared to time their arrival with the temperature reaching about 29° C. This ties in with work by Dudzinski and Arnold (1979) which indicated that sheep could predict that it was to be a hot day and so started their grazing earlier. Further, cattle were found on hot, sultry days to do most grazing in early morning and in the evening and more than usual at night (Weaver and Tomanek, 1951). In hot humid weather like that in Louisiana in summer, the amount of grazing done during the day by Hereford and Aberdeen Angus cows and calves was affected by the shelter available (McDaniel and Roark, 1956). Cattle on pastures without shade showed heat stress that was exhibited as excessive walking, crowding in corners, frequent drinking and lying down in the afternoon when temperatures had fallen. In contrast, the cattle with shade began grazing at approximately the same time that those without it began lying down. Cattle take time to adjust to hot humid weather. In tropical Queensland cattle took three days to acclimatize to the onset of a hot spell of weather (Larkin, 1954).

The periodicities of the major grazing periods of sheep in summer in Southern Australia were analysed in relation to daylength, temperature and humidity (Dudzinski and Arnold, 1979). It was found that the time that morning grazing began was influenced only by the time of sunrise, but that the time it stopped depended on how much grazing had been done the previous night, the time the animals started, and the rate of humidity decrease during the morning grazing period. The start of the afternoon grazing period was influenced by the maximum temperature, the time that it occurred and by the time of sunset. The time afternoon grazing stopped depended on the time the animals started, the time of sunset, the temperature when they started and the maximum temperature.

These findings show that sheep, though insulated by wool, are sensitive to both temperature and humidity and are also sensitive to daylength because they relate the start of the afternoon grazing period to the time of sunset. Different breeds of sheep respond differently to temperature and humidity in summer (Dudzinski and Arnold, 1979). Of the six breeds studied, Dorset Horn and Romney sheep were less sensitive than Suffolk, Southdown, Border Leicester and Cheviot sheep.

Cold conditions appear to have little effect on the diurnal pattern of grazing in sheep. Studies by the authors on the Southern Tablelands of New South Wales (Fig. 1.4) show that the time at which morning grazing began was unaffected by the temperature the previous night, within a range of −7 to 9° C. The only apparent effect of cold nights was that night grazing began earlier and that it was more intermittent than on warmer nights. However, these effects were not pronounced and regression analysis showed that only 35% of the variation in the time at which night grazing began was accounted for by temperatures and the time that the next morning's grazing began.

Cold stress due to shearing may alter the diurnal pattern. In winter, shorn sheep graze more during the day and less at night than unshorn sheep (Hutchinson and McRae, 1969).

Differences between breeds of cattle and sheep

Breed differences in response to temperature have been reported which probably reflect differences in physiological response. Ittner et al. (1954) found that Brahman cattle began grazing earlier in the morning in summer in the Imperial Valley, California and stopped grazing and sought shade later than Shorthorns and Herefords. Santa Gertrudis cattle stayed in the sun even longer.

In another study Aberdeen Angus and Angus × Brahman cattle spent less time grazing during the middle of hot humid days in Louisiana than did pure-bred Brahmans (Rhoad, 1938). Conversely, in winter on rangeland in New Mexico, Santa Gertrudis cattle spent much less time grazing at night

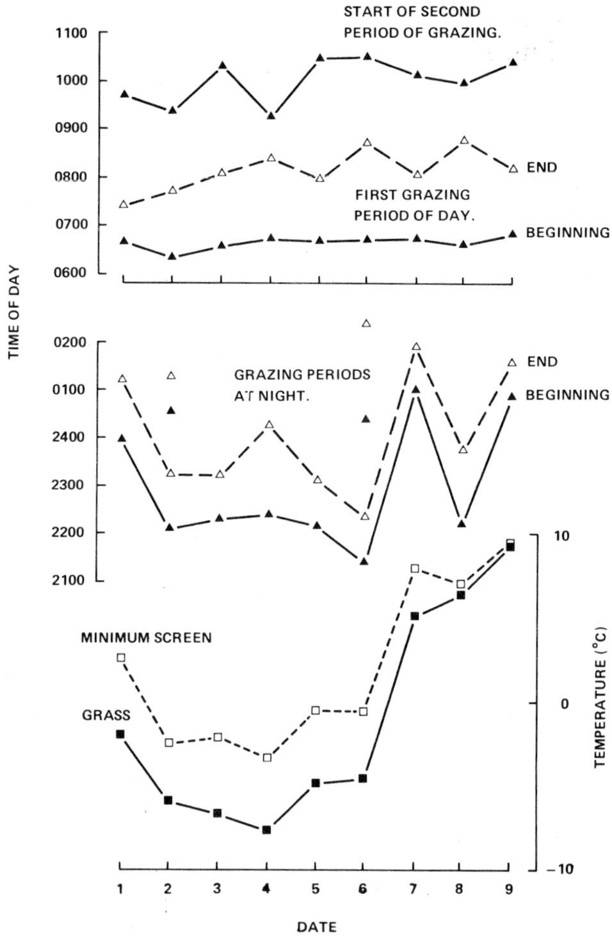

Fig. 1.4. Time periods of grazing at night and in the early morning at Canberra in winter in relation to temperature.

than did Hereford cows (Herbel and Nelson, 1966). Another large difference in the diurnal pattern of grazing of different breeds was found on rangeland in Oklahoma. Hereford heifers spent much less time grazing from midnight to 0600 h in winter and spring than did Hereford × Friesian and Friesian heifers (Kropp et al., 1973).

In very cold conditions in Alberta, cattalo (bison × cattle) were found to graze on more days than Shorthorn, Angus or Hereford cattle (Smoliak and Peters, 1955).

There is some evidence that breeds of sheep differ in their diurnal patterns of grazing. In one study at Canberra, Merino and Border Leicester × Merino sheep were found by the authors to start grazing later in the morning in

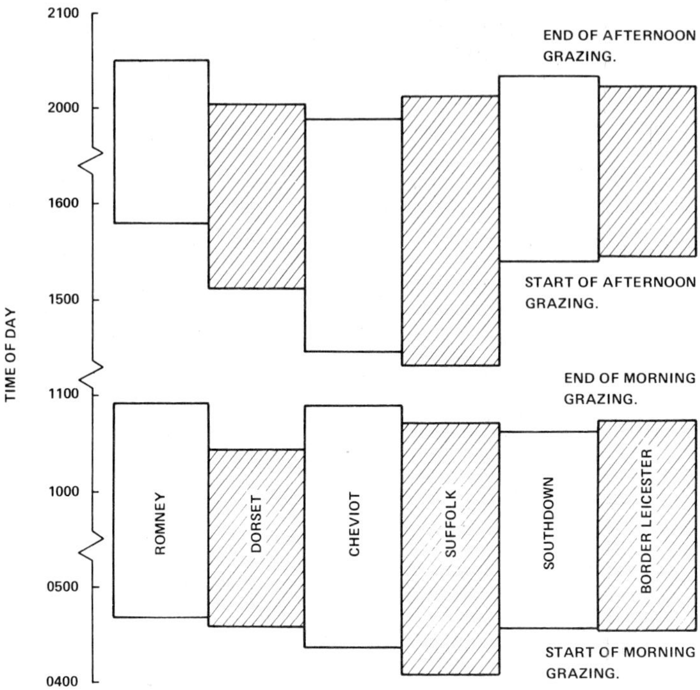

Fig. 1.5. Differences between breeds of sheep in the time of starting and stopping grazing in the morning and at night in summer at Baker's Hill. (Source: Dudzinski and Arnold, 1979.)

autumn than Dorset Horns, with Corriedales beginning the earliest. In another study sheep of five breeds were studied in summer, in a mediterranean climate in Western Australia (Dudzinski and Arnold, 1979). Cheviot and Suffolk sheep differed overall from sheep of four other breeds because they began grazing earlier in both the morning and afternoon than sheep of other breeds (Fig. 1.5). The Romneys differed from the Dorset Horns because they started and finished grazing later in both mornings and afternoons.

The effects of cloud, wind and rain on the diurnal pattern

On days with heavy cloud cattle may delay the start of morning grazing and stop earlier in the evening, cloud effectively reducing daylength (Gonzales, 1964). Also, there is less conformity in behaviour of heifers on overcast days, and the midday break from grazing may disappear (Rutter, 1968). This effect of cloudy conditions on the diurnal pattern of grazing has been observed by the authors with sheep. Cattle are more restless, graze less intensely and cover more ground whilst grazing in cold, stormy or windy weather (Culley, 1938; Rutter, 1968).

Plate 1.1. Cattle standing in driving rain with their necks stretched out.

Rain appears to have little effect on general behaviour unless it is particularly heavy (Waite et al., 1951; Hancock, 1952; Rutter, 1968). Rain squalls cause cattle and sheep to change their direction of grazing from grazing into the wind to grazing away from it. In heavy persistent rain cattle and sheep stop grazing and stand still; cattle stretch their necks out (Plate 1.1). Calves will stop before cows (Hancock and McArthur, 1951). If grazing is interrupted for 0.5 to 1 h the animals may start grazing again despite the heavy rain. Hail will cause cattle to seek shelter. Horses appear to be less readily affected by rain but will stop grazing in driving rain and stand with their tails towards the wind.

The effect of amount of feed available

The amount of herbage or browse available for consumption does not affect the diurnal patterns of grazing as is shown in Fig. 1.6 and Table 1.1 for sheep. The major grazing periods persist, but with a shortage of feed the animals rest less often and so have fewer periods of grazing. When feed was short there was less conformity between individuals in a herd of heifers (Rutter, 1968).

Variation between individuals

Little has been written about the behaviour of individual animals relative to other individuals. With Border Leicester × Merino sheep, Arnold (1962)

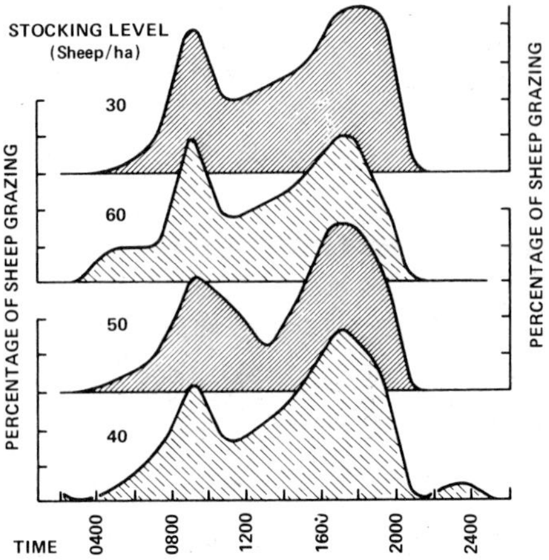

Fig. 1.6. Diurnal patterns of grazing of sheep at different stocking rates. (Source: Arnold, 1962.)

found no evidence that the behaviour of one sheep influenced that of another, even of a twin grazing nearby.

Whilst the beginning and cessation of any major grazing periods vary little between individuals (Fig. 1.7), the variations in between these periods are significant. In a flock of Merino ewes we found that those with longer grazing times stopped grazing significantly later in the first grazing period, started the afternoon grazing period earlier and ended it later.

Most of the above work was done on small flocks of sheep. In a comparison of Merino sheep, North Devon cattle and thoroughbred horses (Arnold, unpublished) it was clear that the different species displayed different degrees of conformity and allelomimetic behaviour when grazing. The sheep acted very much as a unit, the cattle also for much of the day, but the horses did not. This is illustrated for one day in Fig. 1.8; the percentage of the horses grazing at any one time is variable whereas all sheep are either grazing or not

TABLE 1.1

The effect of pasture availability on intensity of grazing (mean of 12 sheep per stocking level) (Source: Arnold, 1962)

Stocking level (sheep/ha)	Number of grazing periods/day	Mean length of grazing periods (h)
30	7.1	0.99
40	5.0	1.73
50	4.7	2.06
60	4.2	2.46

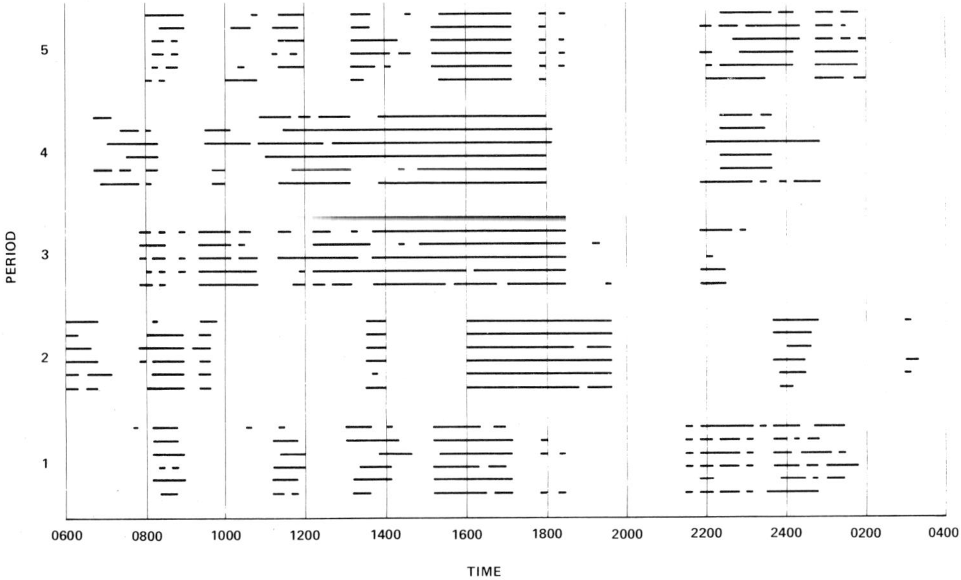

Fig. 1.7. Patterns of grazing for individual sheep in a small flock. (Source: Arnold, 1962.) *Note*: Each line is an individual and the flock was observed on five days.

grazing except for short periods at the beginning and the end of grazing periods. These Merino sheep appear to conform more than the Border Leicester × Merino sheep in Fig. 1.7.

Time spent grazing

The act of grazing involves the selection of herbage, its prehension, mastication and swallowing; the time spent grazing as usually measured includes the time spent searching for food. Grazing time varies in relation to the environment in which an animal finds itself on any given day. The animal varies the number of bites it takes and the size of these bites and both are influenced by the structure of the vegetation. Rate of eating will also be varied according to climatic conditions and the physiological status of the animal. The animal's rate of eating and the time spent grazing determine the amount of food consumed: just how these factors vary in different circumstances is described in detail later. However, the majority of studies report only grazing times, and variations in this will be considered first.

The range in grazing times reported in cattle and sheep is shown in Fig. 1.9. The range for dairy cows is smaller than that for beef cattle and sheep. However, the mean time spent grazing is higher for beef cattle than for dairy cattle or sheep. The slightly higher mean grazing time for beef cattle compared with dairy cattle probably reflects a management effect whereby dairy

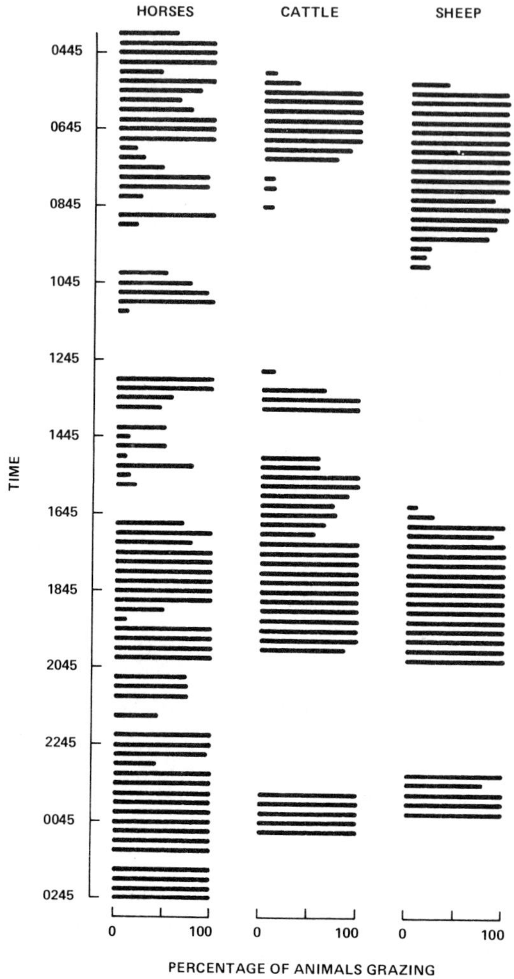

Fig. 1.8. Differences between horses, cattle and sheep in the percentage of animals grazing at the same time.

cows do not graze on pastures as short as those available to beef cattle.

There have been few measurements on other species. Free-ranging camels in the Sahara graze for 8—10 h a day (Gauthier-Pilters, 1974). Goats on rangeland were observed to spend 6.7—9.1 h grazing and browsing, depending on feed conditions (Askins and Turner, 1972). Comparative studies have been rare also. Cory (1927) compared sheep, cattle and goats on adjacent paddocks of rangeland in Texas and obtained grazing times for cattle of 7.7 h, sheep 6.6 h and goats of 5.9 h, which are quite considerable differences. The comparison of sheep, cattle and horses in the same paddock by

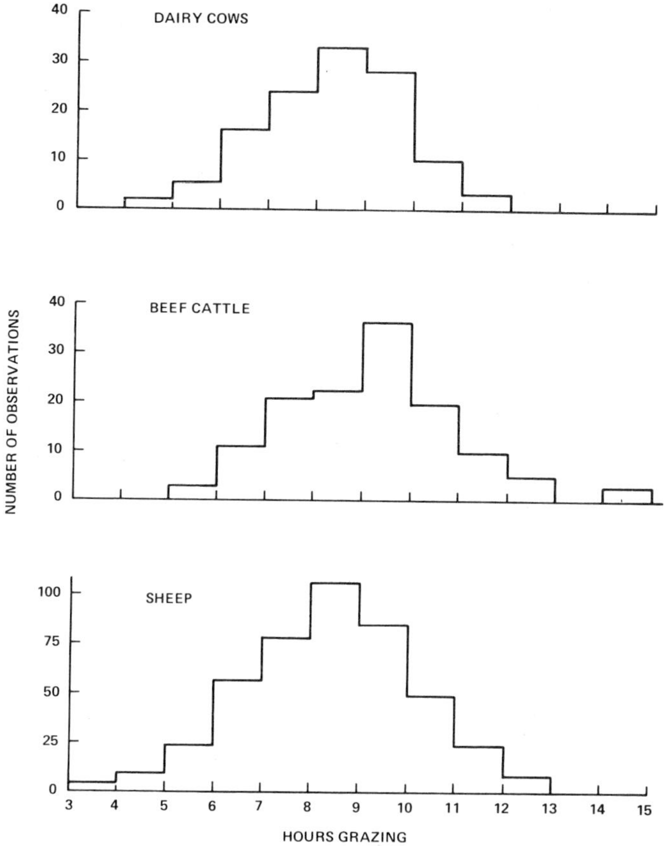

Fig. 1.9. Distribution of times spent grazing by dairy and beef cattle and sheep.

Arnold (unpublished) was over a wide range of forage conditions. With no supplements available, horses grazed for 2 h longer than cattle in winter, and for 6.5 h longer in summer, with sheep having an intermediate grazing time. These results suggest that cattle are more efficient at harvesting their food than sheep or horses, certainly in terms of the energy expenditure in grazing.

The effect of temperature and humidity

Time spent grazing by cattle is affected by very cold, and by hot and humid conditions.

In winter in Alberta with mean temperatures between —8 and —14° C the proportion of cattle grazing increased with temperature (Smoliak and Peters, 1955). Likewise in Utah, Hereford cows spend more time grazing on warm days than on cold days within the range of —24 to 5° C in mean daily air temperature (Malechek and Smith, 1976).

Within the temperature range 0—34° C, temperature could not be related to grazing time in a number of studies if humidity was low (Shepperd et al., 1957; Harker et al., 1961; Wilson, 1961; Gonzales, 1964; Box et al., 1965). In warm humid climates total grazing time has been found to decrease above 26° C (Seath and Miller, 1946, 1947 in Lousiana; Bennie, 1956 in Mauritius). In Oklahoma, Dwyer (1961) found that on hot days grazing time was reduced, there being no compensatory increase in night grazing time. Grazing time was equal to

$$8.12 - 0.07 \times \text{average temperature}$$

in the temperature range 21—32° C. In Missouri, the grazing time of cattle in summer was more closely related to a comfort index used for humans (Ehrenreich and Bjugstad, 1966).

$$\text{Temperature-humidity index} = 0.4\ (\text{total dry bulb} + \text{wet bulb temperatures}) + 15$$

They also found that on hot humid days grazing time increased with wind speed.

In Alberta cattle and cattalo decreased their grazing time as wind speed increased, and in Utah a change in barometric pressure was followed by an increase in grazing time.

Relatively little is known about the effects of climate on grazing time in sheep. Sheep provided with shade in summer in Canberra grazed for 10.4 h compared with 9.0 h for sheep without shade (Arnold, unpublished). This may indicate that the sheep found it more comfortable to graze than to rest in hot conditions when no shade was available because the sheep with shade had no shaded areas in which to graze.

Differences between breeds within species

In cattle, small differences between breeds in grazing time per day have been reported. The greatest was 1.4 h between Hereford and Santa Gertrudis cattle grazing on range in New Mexico (Herbel and Nelson, 1966). The difference between Hereford, Hereford × Friesian and Friesian heifers was only 39 min (Kropp et al., 1973). Both studies were year round. Over shorter periods Brumby (1959) found no difference in grazing times of lactating Jersey and Friesian cows in New Zealand. In Uganda, Friesian steers grazed for 1.6 h longer than Nganda steers (Musangi, 1965), but the Nganda steers were lighter in weight and grazing time per kilogram of liveweight was similar. If this effect held generally then Brumby should have obtained a large difference between the Jerseys and Friesians.

Brumby used regression analysis to compare grazing times adjusted for weight of the cows and their milk yield but could still show no difference between the breeds. The problem is that grazing time is not the sole index of effort in grazing. Possibly the Jerseys ate more slowly and so grazed for the same time as the Friesians. On the other hand, social facilitation between the two breeds could have reduced any differences because they were grazed together. This aspect is discussed later.

There are few comparative studies with sheep. Arnold and Dudzinski (1967a) found in separate studies that Corriedale ewes had shorter grazing times than Border Leicester × Merino ewes when dry, pregnant or lactating. However, Corriedale sheep ate much less than Border Leicester × Merino sheep. In a later study the grazing times for four breeds of sheep varied by as little as 5% or as much as 15% depending on the pasture conditions and the differences in rate of eating (Arnold, 1975). These interrelationships are discussed later. However, over 21 observation periods of a week each throughout the year, the means were Dorset Horn — 9.8 h, Border Leicester × Merino — 9.5 h, Merino — 9.5 h and Corriedale — 9.1 h. In extreme climates native breeds may be better adapted physiologically. An example is that for sheep in Egypt (Sharafeldin and Shafie, 1965) which are grazed during the day only. The local Ossimi sheep grazed in summer for 4 h whereas exotic breeds did so for only 2 h because of heat stress.

Development of grazing in the young

Chambers (1959) followed the development of grazing time in bucket-reared calves. The average time spent grazing increased rapidly from 10 min at a week old to 4 h at 3 weeks and 5.5 h at 2 months. With single suckle cows, the calves grazing time was as shown in Fig. 1.10. Calves did not graze for any significant period until they were 20 days old and then grazing increased to 74 days of age when the study ceased. Hutchinson et al. (1962) studied calves up to 6 months of age, obtaining very similar results.

Calves that are 4—6 months old spend 18% less time grazing than their dams (Peterson and Woolfolk, 1955; McDaniel and Roark, 1956; Dwyer,

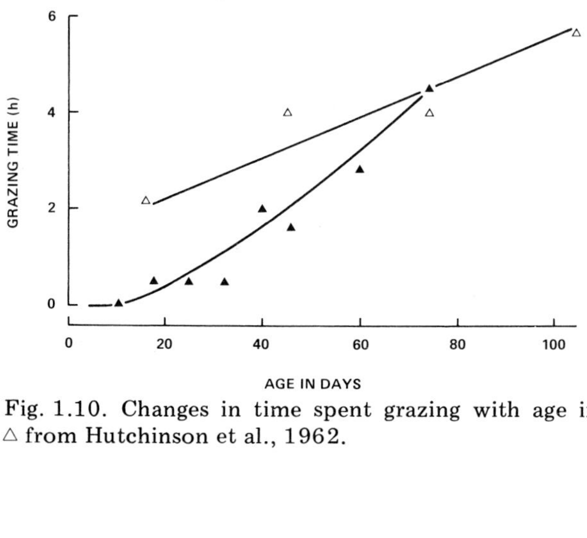

Fig. 1.10. Changes in time spent grazing with age in calves. ▲ From Chambers, 1959; △ from Hutchinson et al., 1962.

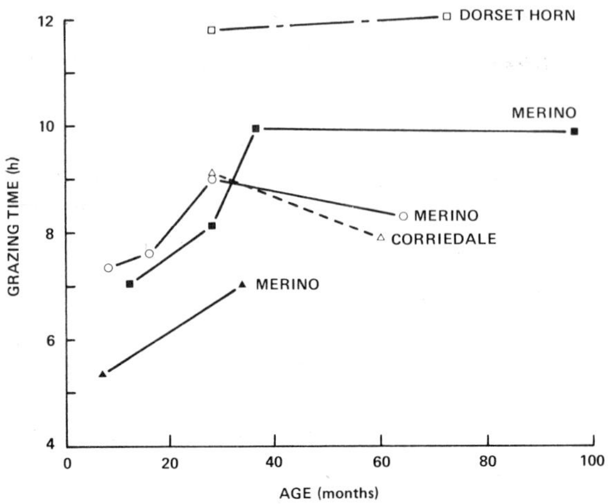

Fig. 1.11. Changes in time spent grazing with age in sheep.

1961). Hodgson and Wilkinson (1967), using Jerseys, found that 8—17 month old calves grazed for longer (8.5 h) than 17—24 month old heifers (7.4 h) or $3\frac{1}{2}$ year old cows (6.9 h). Thus by a year of age calves will graze for at least as long as mature cows.

There are no published data for sheep. However, the authors have found that grazing time increases with age up to about 3 years (Fig. 1.11). After that, grazing time may decline. A decline with age may occur because the teeth become worn or too long, but there is no evidence on this.

The effect of physiological condition

The time spent grazing reflects part of the animal's response to its needs. Two drives are important and can influence grazing time. These are the need for food and the need for shelter. The animal will have an increased food requirement if it is pregnant or lactating, cold or has experienced a period of poor nutrition. If it is cold it will seek shelter and may have to adjust its grazing behaviour to do so. The extent to which the animal can meet an increased food requirement will, in part, be determined by the availability of herbage and the ease with which it can be harvested. Nearly all the work in this area has been with sheep.

Table 1.2 shows the mean grazing times of four breeds of ewes at four levels of pasture. The ewes all responded in the same way. Pregnant ewes grazed for the same time as dry ewes but lactating ewes spent 12% more time grazing than dry ewes at all but the highest level of pasture available.

There seems to be some consistency in the response; two earlier experiments by the authors (Table 1.3) gave similar results. In each there was a range of pasture conditions: both Border Leicester × Merino and Corriedale ewes grazed for a little longer in late pregnancy than did dry ewes. In early

TABLE 1.2

Grazing times of ewes of four breeds at different physiological states (Source: Arnold, 1975)

Pasture level	Pregnancy		Lactation	
	Pregnant ewes	Dry ewes	Lactating ewes	Dry ewes
High	8.6	8.5	8.9	9.6
Medium	8.5	9.4	9.0	7.7
Low	10.2	10.6	9.5	8.4
Very low	10.7	10.7	10.0	9.0
Mean	9.5	9.8	9.4	8.7

lactation Corriedale ewes had a 7% higher grazing time and Border Leicester × Merinos a 12% higher grazing time than dry ewes.

In general, thin sheep eat more than fat sheep under all pasture conditions. They do so because they graze for a little longer than fat sheep but they also eat faster (Arnold and Birrell, 1977). The results of several experiments are shown in Table 1.4. Cold stress following shearing results in a decrease in grazing time and sheep obtain more food when cold by eating much faster. In two experiments at Hamilton (Arnold and Birrell, 1977), the shorn sheep increased their rate of eating by 43%. Under more extreme conditions, shorn sheep in very poor condition will almost cease grazing on cold, wet days (Fig. 1.12). Shelter when grazing was provided by tall dry grass at the lowest stocking rate, but not at all at the higher stocking rates; this meant that grazing time was little affected by shearing at the low stocking rate but considerably at the high stocking rate.

Unfortunately there have been only two studies on the effects of level of production on grazing time. In both, the grazing time of cows increased significantly with yield of 4% fat corrected milk (Hancock, 1954a; Brumby, 1959). The correlation was small in Brumby's study but in Hancock's work

TABLE 1.3

Grazing times of ewes of two breeds at different physiological states (Source: Arnold and Dudzinski, 1967a)

Breed	Pregnancy		Lactation	
	Pregnant ewes	Dry ewes	Lactating ewes	Dry ewes
Border Leicester × Merino	8.8	8.5	11.2	9.9
Corriedale	7.4	6.8	7.9	7.3

TABLE 1.4

Grazing times of thin and fat sheep on scant and abundant pasture (Source: Arnold and Birrell, 1977)

Experiment	Abundant pasture		Scant pasture	
	Thin	Fat	Thin	Fat
1	5.6	5.7	8.3	7.2
2	7.4	5.0	7.8	7.3
3	6.8	6.7	7.1	6.5
4	8.5	7.6	9.3	9.3
Mean	7.1	6.3	8.1	7.6

55—84% of the variation in milk yield was associated with variation in grazing time.

There have been no specific studies of how size of animal affects grazing time. Brumby (1959) found that when grazing times were adjusted for level of milk production, there was no relationship with the weight of the cows.

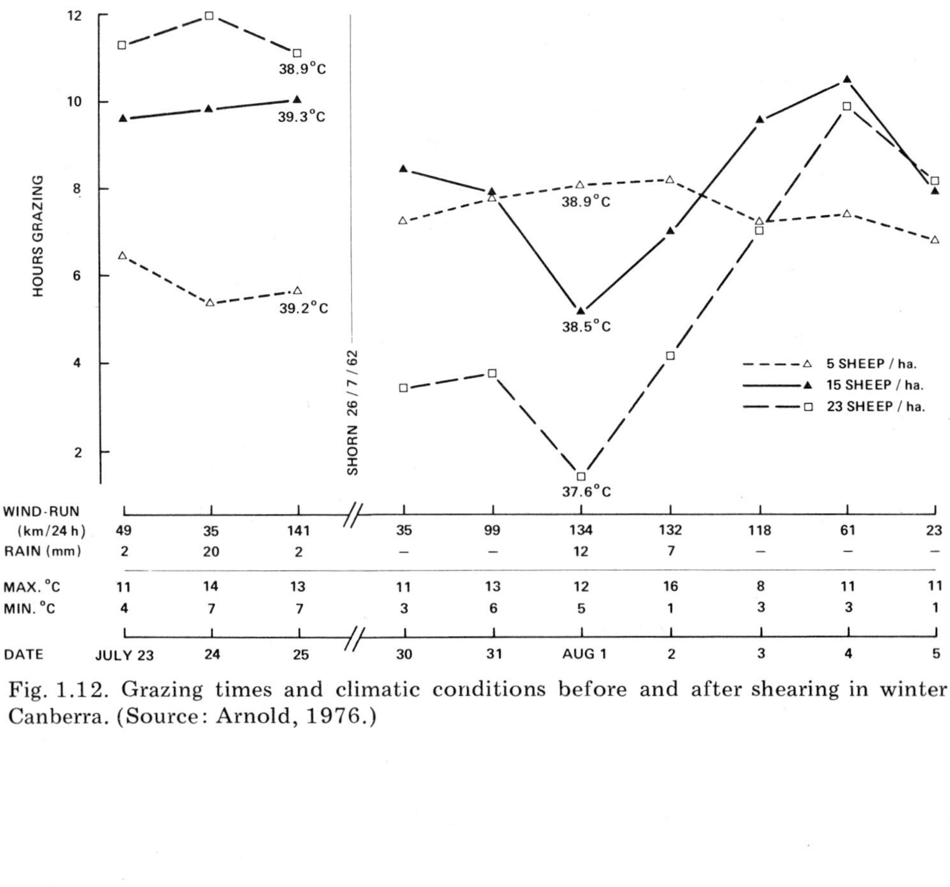

Fig. 1.12. Grazing times and climatic conditions before and after shearing in winter at Canberra. (Source: Arnold, 1976.)

No other analyses have been done for cattle. A small study (Arnold, unpublished) showed that, for sheep, jaw size and age are more important than liveweight in determining differences between sheep in grazing times. However, these three factors accounted for only 29% of the variation in time between individuals.

The effect of social facilitation

Groups of animals whose potential grazing times differ may interact socially if grazed together, with the result that the potential grazing times are modified. There have been several experimental studies of this phenomenon, which is termed social facilitation.

Tribe (1950) found that ewes fed a supplement grazed for an hour longer than unsupplemented ewes when grazing alone, but not when they grazed with the unsupplemented ewes. Subsequent work with sheep (Holder, 1962) gave variable results but in each of three experiments supplemented sheep grazed for a longer period and unsupplemented sheep for a shorter period when grazed together than when grazed alone. With cattle, however, no such effects have been found (Bailey et al., 1974).

Many flocks and herds contain animals of several different ages, and the grazing times of particular age groups can be different if the animals are grazed separately. We found that the differences in grazing times of Merino wethers of five ages were much greater when they were grazed separately than when grazed together. A similar effect has been obtained with cattle (Hodgson and Wilkinson, 1967). In another study we found that the difference in grazing time between lactating and dry Merino and Dorset Horn ewes was 16% greater when they grazed separately than when they grazed together.

It must be concluded that social facilitation does influence the grazing time of groups of sheep. But it is difficult to understand why this occurs. The individuals within a flock may vary in grazing time from, say, 8 to 12 h a day. This variation occurs because of individual differences in the need to graze. Social facilitation between individuals occurs which results in the beginning and cessation of major grazing periods varying relatively little between sheep (Fig. 1.7). However, we have found that sheep with longer grazing times stop grazing later in these main periods and in the afternoon start earlier as well. But most of the variation in grazing time is due largely to variation in the amount of time spent actually grazing within these periods. Because of this, it would be expected that social facilitation between groups of sheep having different needs to graze would not occur because the individuals would express the normal differences as usual. Obviously, a more detailed analysis is needed to find out how the effect operates.

The effect of supplementary feed

The grazing time of cattle is decreased when more than a small supplement of concentrate is fed but not necessarily when hay is fed. Feeding 0.7 kg of cottonseed to Herefords on range in winter did not affect grazing time (Box et al., 1965) but feeding 3.6 kg of oats reduced grazing time of Friesian steers by 2 h (Bailey et al., 1974). On high quality young pasture Musangi (1966) found that feeding 3.6 kg barley reduced grazing time of steers by 19—28%.

Concentrate supplements fed to dairy cows in proportion to their milk production can reduce grazing time by as little as 10% (Hardison et al., 1956; Holder, 1960) to as much as 58% (when fed at 0.33 kg per kg of milk — Moseley, 1929). However, Seath and Miller (1947) found that feeding 2 kg of hay to cows on good pasture hardly affected their grazing time, a result similar to that with steers mentioned earlier.

In the comparison of sheep, cattle and horses mentioned earlier there were periods during which supplements were fed in both winter and summer (Table 1.5). The supplements caused much greater reductions in the grazing time of cattle and horses, than in that of sheep.

The effect of forage availability

It can be generally said that when feed is short on either rangelands or pasture then the animals grazing there will spend more time grazing because they can obtain less per bite. Even though rate of biting is increased with decreased pasture height the point is reached where the animals cannot compensate and food intake decreases (Arnold and Dudzinski, 1969). The interrelationships between grazing time, bites per minute and amount per bite vary with the physical structure of the pasture or range forage — its bulk, its density and its height. These interrelationships have been discussed by Arnold and Dudzinski (1967b) and Allden and Whittaker (1970) for sheep, and generalised and specific relationships are shown in Figs. 1.13 and 1.14. At a given height the spatial arrangement over a paddock of ryegrass was un-

TABLE 1.5

Grazing times of sheep, cattle and horses grazing together (h)

	No supplements available	Supplements
Winter — green feed		
Sheep	10.4	9.6
Cattle	9.5	5.1
Horses	11.5	5.2
Summer — dry feed		
Sheep	9.5	6.7
Cattle	7.7	3.0
Horses	14.2	7.2

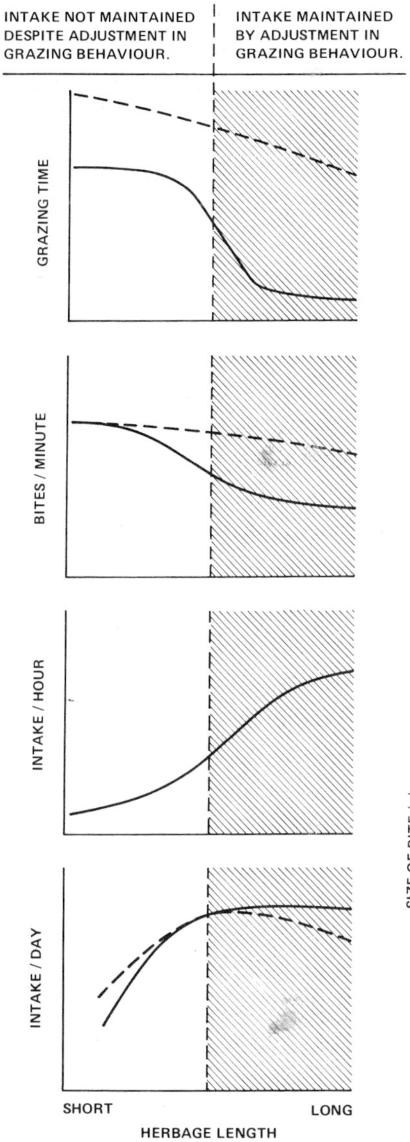

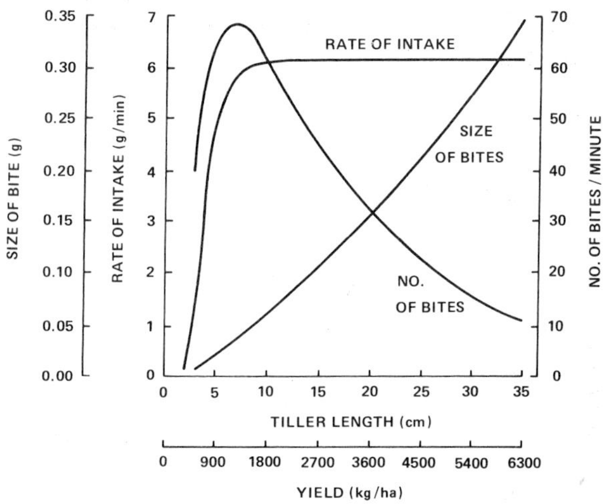

Fig. 1.13 (left). Changes in the grazing behaviour of sheep with changing pasture conditions.

Fig. 1.14 (right). Relationships between tiller length and number of bites, size of bites and resultant intake of grazing sheep. (Source: Allden and Whittaker, 1970.)

important to sheep so long as the density where pasture was growing was similar (Allden and Whittaker, 1970). Arnold and Dudzinski (1969) analysed the effects of the density and showed that a a given height of pasture food intake could vary by 50% because of differences in density. In other words, the lower the density, the less eaten per bite. There is no information on whether the number of bites can be increased at a low density, but circumstantial evidence suggests that this is so. For example, Allden and Whittaker obtained biting rates between 10 and 70 per minute on ryegrass, with the value being 70 when tiller length was 5 cm and yield 900 kg ha^{-1}. Recent studies by the authors give values 50% higher at the same pasture yield but the density was lower.

For sheep the changes in grazing behaviour with changing pasture conditions cannot be predicted. Responses differ for no known reason. Thus in one experiment (Arnold, 1975) grazing time differed markedly with pasture yield over one period of 4 weeks, but a month later showed virtually no change; rate of eating followed the reverse pattern. Thus the relationship between grazing time and rate of intake can vary as shown in Fig. 1.15. An analysis of data for different breeds of sheep (Arnold and Dudzinski, 1967a; Arnold, 1975) shows that differences in rate of eating are more readily predicted. Thus the factors that determine time spent grazing by sheep are at present poorly understood.

There have been fewer attempts to analyse the interactions between grazing behaviour and pasture structure in cattle. Stobbs (1974) compared the grazing behaviour of cows on temperate and tropical pastures. Even on young herbage cows took fewer bites per minute (75—100) on temperate pastures than on tropical pastures (85—150) and grazed for less time. On

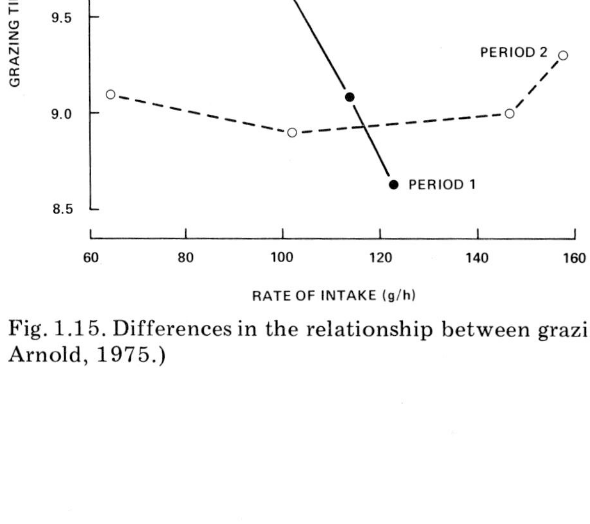

Fig. 1.15. Differences in the relationship between grazing time and rate of intake. (Source: Arnold, 1975.)

mature tropical pastures which are up to a metre in height but have a low bulk density, bites per minute were as high as 290 and grazing times exceeded 13 h a day. Other values for bites per minute on temperate pastures range from 16—44 (Hodgson and Wilkinson, 1967, 1968) to 44—70 (Hancock, 1950, 1954a). Hancock found that the rate of biting varied during the grazing period from 60—70 a minute at the beginning to 30—50 at the end. On green annual pastures in Western Australia we have found values between 77 and 102, depending on pasture yield (Fig. 1.16). Horses, by comparison had values of 66—93.

Changes in grazing behaviour of cows as they progressively defoliated a *Setaria anceps* pasture in autumn and spring have been studied (Charcon and Stobbs, 1976). The initial amounts of herbage present were 7400 kg ha^{-1} in autumn and 3800 kg ha^{-1} in spring. Even when the cows began with more than 7000 kg ha^{-1}, food intake declined continuously as they defoliated the sward (Fig. 1.17). On day 14 in autumn there was still 3200 kg ha^{-1} of total herbage, yet intake had fallen to 25% of its initial value. Rate of biting changed very little, compared with studies on other types of tropical pasture (Stobbs, 1973b), and although grazing time increased, bite size decreased rapidly. Apparently the cows were being very selective about the parts of the plants they would eat, to the detriment of food intake. Thus in this situation changes in grazing behaviour at no time compensated for changes in pasture yield and composition to maintain food intake. Selective grazing and its effect on food intake are discussed in detail in Chapter 3.

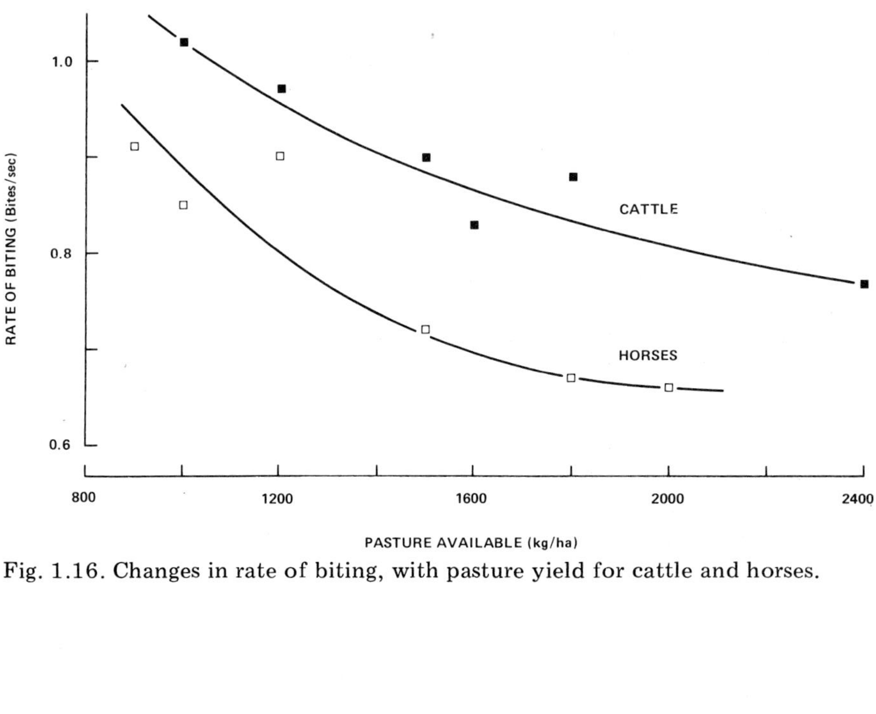

Fig. 1.16. Changes in rate of biting, with pasture yield for cattle and horses.

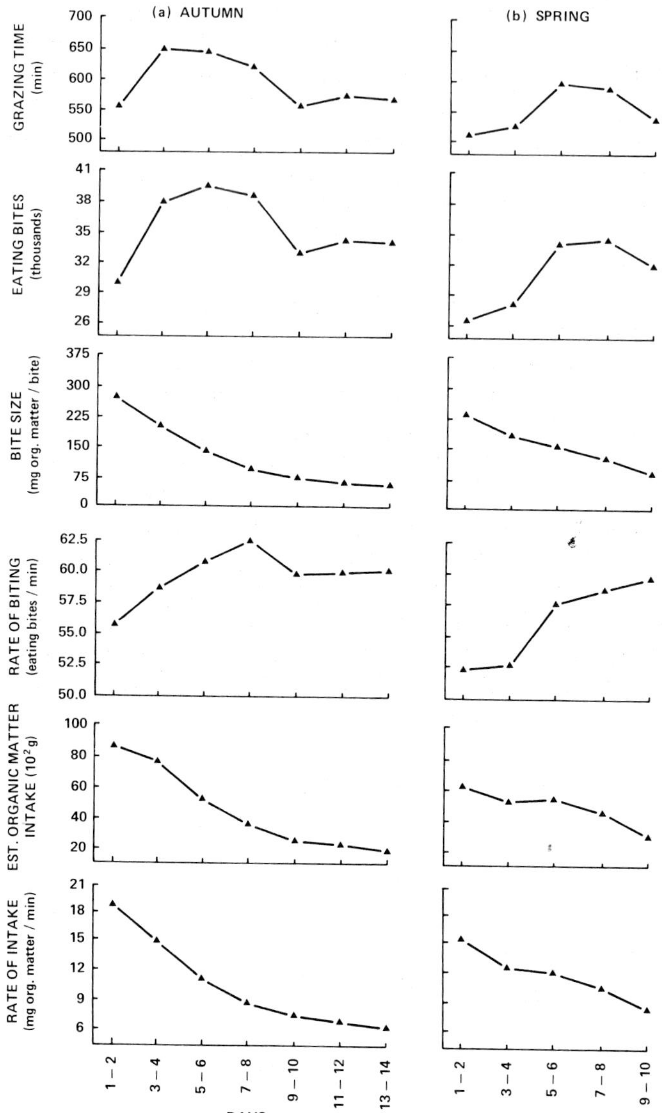

Fig. 1.17. Variation in grazing behaviour of cattle grazing Kazungula *Setaria* pastures at different stages of defoliation (a) in autumn and (b) in spring.

On another type of tropical pasture cows were found to reduce grazing times drastically when pasture yield fell below 2500 kg ha^{-1} (Cowan and O'Grady, 1976). No reason can be given for this unusual finding — more detailed measurements such as those made by Stobbs would be needed before the finding could be explained.

RUMINATION

Chewing the cud is the second most time consuming activity of ruminant animals. Fig. 1.18 shows that the time cattle spend ruminating varies from 1.5 to 10.5 h a day, but in the majority of published studies it is between 5 and 9 h. There have been too few studies with sheep to present a histogram: the range is the same but the mean time spent ruminating seems to be shorter with sheep. Thus Lofgreen et al. (1957) found that the ratio of ruminating to grazing time was 0.95 in steers and 0.46 in sheep grazing the same pasture. Cory (1927) recorded values of 76 min for cattle, 78 for sheep and 84 for goats; these values are very low and must be doubtful.

The time taken depends on the quantity and quality of food eaten and the amount of 'grinding' it requires. In New Zealand Hancock (1954b) found that the ruminating time of cows increased from 6.0 h to 8.2 h as the rye-grass—white clover pasture changed from the vegetative to full flowering stage of growth. Similarly, cows grazing rangeland in Central Australia spent 5 h ruminating in winter on good quality feed and 8 h in summer on poor quality feed (Low et al., 1978b, submitted for publication). Since coarse mature herbage is less digestible than succulent young herbage, less of it is eaten; so most studies tend to show only small seasonal change in ruminating time (Herbel and Nelson, 1966; Kropp et al., 1973). However, there have been no critical studies of the factors that influence ruminating time in grazing animals.

The difference between cattle and sheep may be because sheep selectively eat less fibrous herbage (see Chapter 3), they take smaller bites and the material when swallowed is finer, so presumably needs less chewing during rumination.

There is no distinct, diurnal pattern in rumination as there is in grazing, because the time when it occurs is largely influenced by the grazing pattern. Since relatively little grazing is done at night, much of the ruminating takes place then. Five different studies with cattle show a remarkable consistency

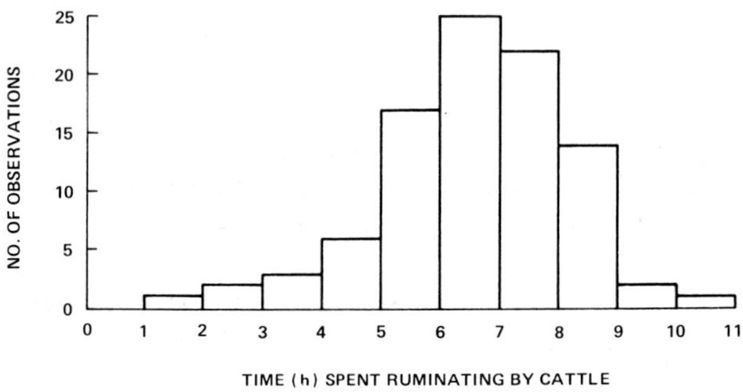

Fig. 1.18. Distribution of times spent ruminating by cattle.

in the percentage recorded between 0600 and 1800 h (64—71%).

Most ruminating is done when the cows are lying down; the range reported is between 62 and 83%. Wagnon (1963) carried out one of the most intensive studies with Hereford cattle, and found that 80% of rumination was done when the cows were lying down, 17% when they were standing and 3% when they were nursing their calves. The length of a rumination period was found to vary between 1 and 100 min, with an average of 34 min.

Little information is available for sheep. In 20 observations with Border Leicester × Merino sheep there was an average of nine ruminating periods a day with an average length of 22 min (Arnold, 1962). It was also found that ruminating time decreased with stocking rate (Arnold, 1960a), presumably because sheep at higher stocking rates are usually eating younger, more succulent material that requires less mastication.

Differences between breeds of cattle are small. Kropp et al. (1973) found none between Herefords, Friesians and Hereford × Friesian heifers. Herbel and Nelson (1966) found small differences between Hereford and Santa Gertrudis cows, with the Santa Gertrudis cows doing a little less ruminating at night but more of their ruminating whilst lying down (86% compared with 79% for Herefords).

Supplementary feeding of cattle may influence the time spent ruminating. Range cows fed 0.7 kg cottonseed cake a day in winter ruminated for 5.1 h compared with 2.1 h for unsupplemented cows even though grazing times differed by only 14 min (Box et al., 1965). On the other hand Wagnon (1963) reported no difference in ruminating time of supplemented and unsupplemented cows. He did find that the number of chews per bolus decreased from 47 to 39 when supplements were provided. The results of Box et al. suggest that the supplement was having a marked effect on rumen stasis.

It appears that the greatest effect of forage maturity is to change the number of chews per bolus. Wagnon obtained a range from 32 on young forage to 45 on mature forage. Calves chew faster but have more chews per bolus (55 compared with 49 for cows — Dwyer, 1961).

RESTING

The proportion of resting time spent lying down varies between species of animals. As can be seen from Table 1.6, sheep rest more frequently by lying down than do cattle, whereas horses spend only a small proportion of their resting time lying down.

Nearly all published data refer to cattle and in these the proportion varies from 0.55 (Dwyer, 1961) to 0.83 (Wagnon, 1963), both values being for Hereford cows. There appears to be no clear cut seasonal pattern, as the two contrasting sets of figures in Table 1.7 show.

TABLE 1.6

Comparison of proportion of resting time spent lying by sheep, cattle and horses

	Winter	Summer
Sheep	0.88	0.83
Cattle	0.83	0.67
Horses	0.01	0.07

TABLE 1.7

Changes in percentage of resting time spent lying with season

	Autumn	Winter	Spring	Summer
New Mexico (Herbel and Nelson, 1966)	61	75	72	67
Oklahoma (Kropp et al., 1973)	84	70	71	77

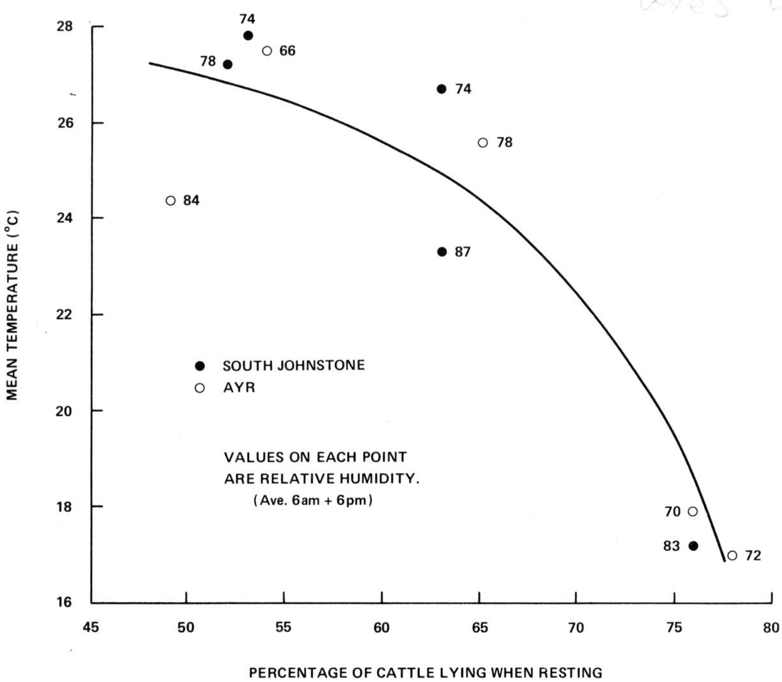

Fig. 1.19. Changes in the percentage of cattle lying when resting with mean daily temperature. (After Larkin, 1954.)

Part of the lack of consistency relates to the time of day when the animals are resting. At night most rest will be taken lying down but during the day climatic conditions have a strong influence. On hot days cows stand more than lie down, especially when there is very little wind (Dwyer, 1961).

Humidity is also important, and physiologically cattle show increasing stress at high levels of humidity when the temperature goes above 28°C. Larkin (1954) studied beef cattle in two locations in North Queensland, which is in the dry tropical region. At both locations the proportion of resting time spent lying down decreased with increased temperature (Fig. 1.19), and especially when humidity was high. At one centre, South Johnstone, the use of shade was recorded. Fig. 1.20 shows quite clearly that the use of shade increased rapidly with increased temperature. In Florida, Aberdeen Angus cattle always rested in shade during the day, but Brahman cattle did not (Rhoad, 1938). The crossbreds of these two breeds were intermediate in behaviour (Fig. 1.21).

In very cold conditions cattle stand less with increasing wind speed, presumably to reduce heat loss (Malechek and Smith, 1976). Under milder winter conditions cattle spend a greater proportion of their resting time standing than in summer (Rutter, 1968); again this would seem to be to

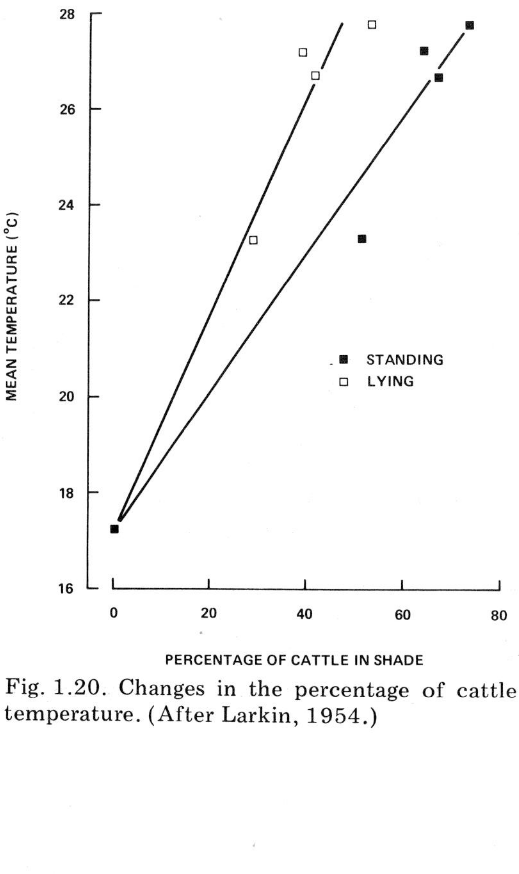

Fig. 1.20. Changes in the percentage of cattle using shade for resting with mean daily temperature. (After Larkin, 1954.)

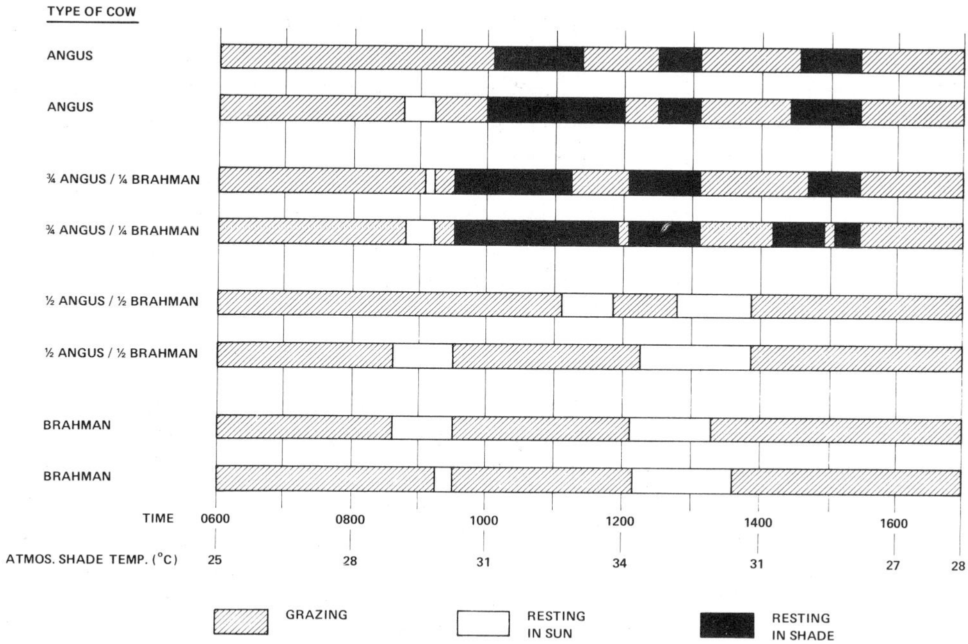

Fig. 1.21. Differences between breeds of cattle in grazing and resting habits. (Source: Rhoad, 1938).

minimise heat loss from lying on cold wet ground, but comfort may have been involved.

There is less information available for sheep. At Stratford, England, Hughes and Reid (1952) found that the proportion of resting done lying down varied from 0.6 to 0.9 but did not say how weather conditions influenced these values. In very hot conditions sheep with no access to shade will often stand in tight groups rather than lie down. This feature is discussed later under shelter-seeking behaviour. Recently shorn sheep spend more time standing than unshorn sheep in cold weather, which would have an adverse effect on their thermoregulation and so is an aberrant form of behaviour (Hutchinson and McRae, 1969).

Horses, as has been seen, have a less diurnal pattern of grazing than sheep and cattle and so may have more flexibility in the timing of resting. The time of peak resting by Exmoor ponies depends on weather conditions (Tyler, 1972). When it is overcast all day there is no peak. On other days the ponies rest when the sun breaks through and in summer they rest in shade for up to 5 h a day. When resting in the sun, the animals' bodies are at right angles to the sun's rays. Rutter (1968) noticed that in winter cattle persistently stand at right angles to the sun, but turn downwind when the wind rises; thus they appear to maximise radiant heat exchange and reduce convective heat exchange. In the Sahara, resting camels always face the sun and remain all day on the same spot, but adjust their position to the moving sun. In this way they expose as little as possible of the body surface to the sun (Gauthier-Pilters, 1974). If they had to change places during the day they would get a big influx of heat through their bellies.

The sites chosen for camping are influenced by the physical environment, the weather and by where the animals have been grazing: they may be chosen because they provide shelter from cold or heat. Animals may also seek shelter from both heat and cold whilst grazing.

Location of camping sites

When light lasts longer than about 10 h sheep and cattle usually spend some time lying down during the day as well as at night. Very often the sites chosen for day camping differ from those selected at night. In mild weather the animals camp in the open and there is little evidence to indicate the factors that influence a choice of site. Very often it will be simply at the point where the morning grazing ceases, but in cold, windy weather it is likely to be in the less exposed areas of the paddock in which they are grazing. Fig. 1.22 shows the locations where sheep, cattle and horses camped during winter and summer in a Mediterranean-type area of Western Australia (Arnold, unpublished). It can be seen that locations were widely scattered over a period of months, but differed for the three species. At night in both winter and summer, camping was usually on higher ground. During the day, camping in winter was in areas sheltered from prevailing winds whereas in summer it was primarily in shade, especially for cattle and horses. The three species kept separate whilst camping, although they were often in adjacent areas.

In winter, at night, sheep often choose camping sites in warmer parts of the paddock. This is illustrated (Fig. 1.23) for two flocks observed on the Southern Tablelands of New South Wales by Arnold (unpublished). It is clear that as the nights got colder the sheep of both breeds moved higher up the slopes.

In summer, camping during the day is most often in shade if this is available, as is shown on Plate 1.2. On a Utah range sheep sought shade in the bottoms of canyons when the temperature exceeded 14°C (Bowns, 1971). When there is no shade, sheep will camp on bare ground or dig holes to lie in (Plate 1.3, 1.4) if the ground is soft enough. This is a common habit in Bighorn sheep. The use of any particular camping hole was found to be random and not the prerogative or a particular sheep (Arnold, unpublished). Under very hot and shadeless conditions sheep gather into small groups and shade their heads under each other rather than lie down (Plate 1.5). Camels huddle together, but when lying down (Gauthier-Pilters, 1974).

Both cattle and sheep without shade camp during the day near water

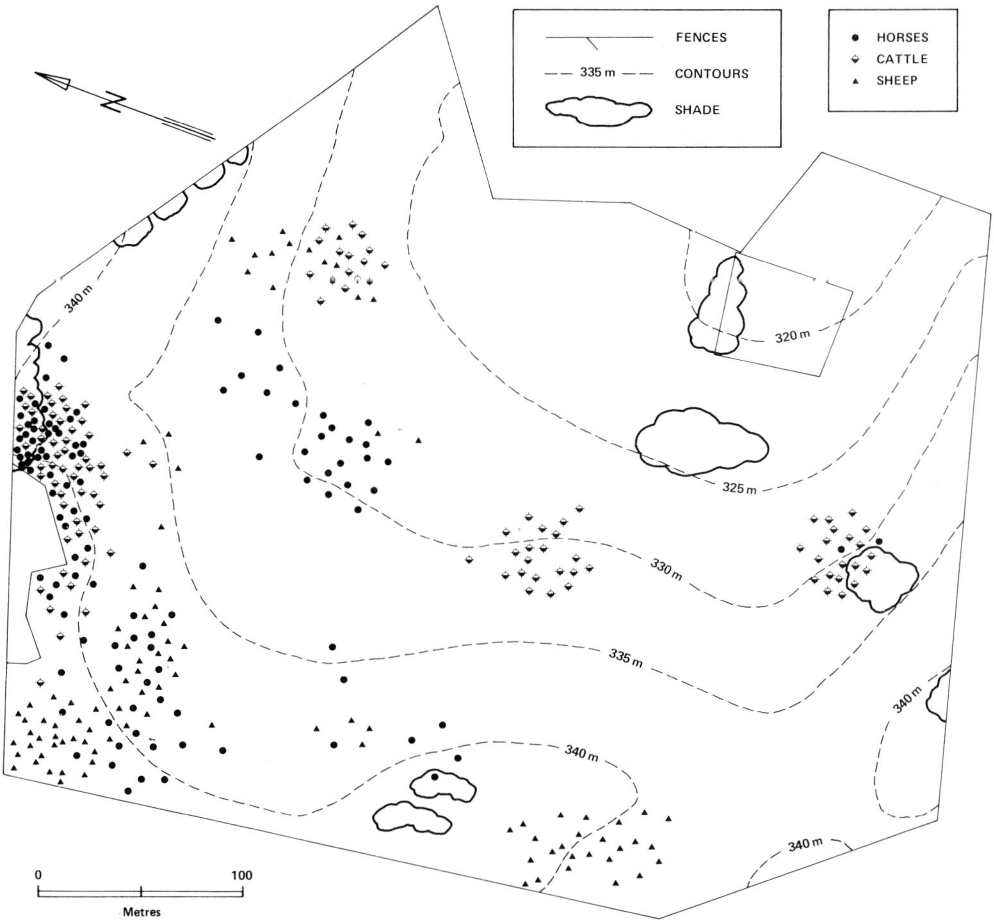

Fig. 1.22. Locations of camping sites of sheep, cattle and horses. (a) Resting at night in winter.

(Weaver and Tomanek, 1951; Schmidt, 1969), but when shade is available it will be used in preference (Schmidt, 1969; McIlvain and Shoop, 1971).

At night in summer sheep camping sites tend to be on high ground, if there is a choice. Bowns (1971) found that Rambouillet and Targhees sheep camped on a few spots on high ridges in the mountains of Utah, using the same ones throughout summer — from warm July nights to the cool nights of September; Columbia sheep camped lower down. In the Wallows National Forest in Oregon Merino and Rambouillet sheep used 77 camping sites in 99 days, with preference being for three sites all on a prominent open point adjoining heavy timber (Jardine, 1910). Weaver and Tomanek (1951) found that cattle camped at night at three preferred sites on rangeland in Nebraska; all were on north (sun) facing slopes with abundant forage.

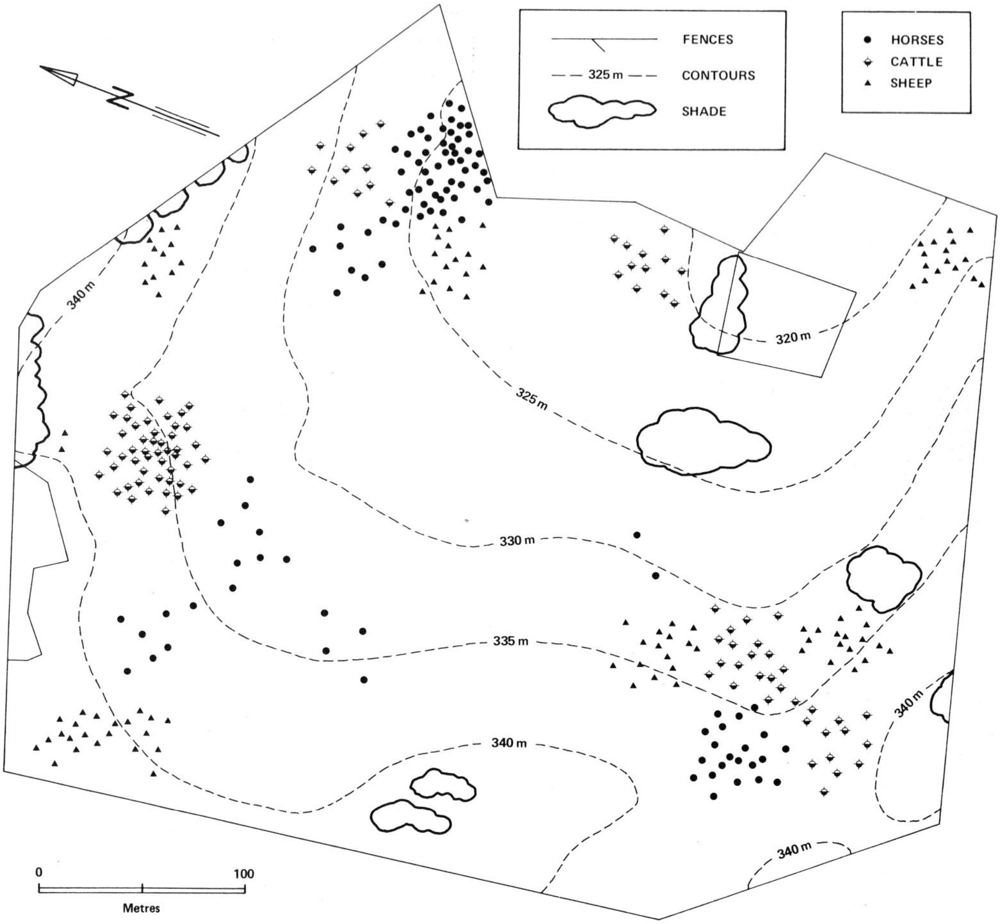

Fig. 1.22. (b) Resting at night in summer.

Use of shade and shelter

Seeking shade is an obvious response to heat stress and this is a function of both temperature and humidity as well as genotype. Temperate breeds of cattle in North Queensland spent up to 11 h a day in the shade (Larkin, 1954). The availability of shade in hot humid climates can affect productivity, as found by McDaniel and Roark (1956), McIlvain and Shoop (1971) and Peacock et al. (1965) in Louisiana, Oklahoma and Florida, respectively. However, shade from artificial shelter is not always as effective as natural shade (McDaniel and Roark, 1956). Artificial shelters need to allow enough space for animals to keep their normal social distance when lying or standing, and, as shelter from heat, to allow maximum air movement.

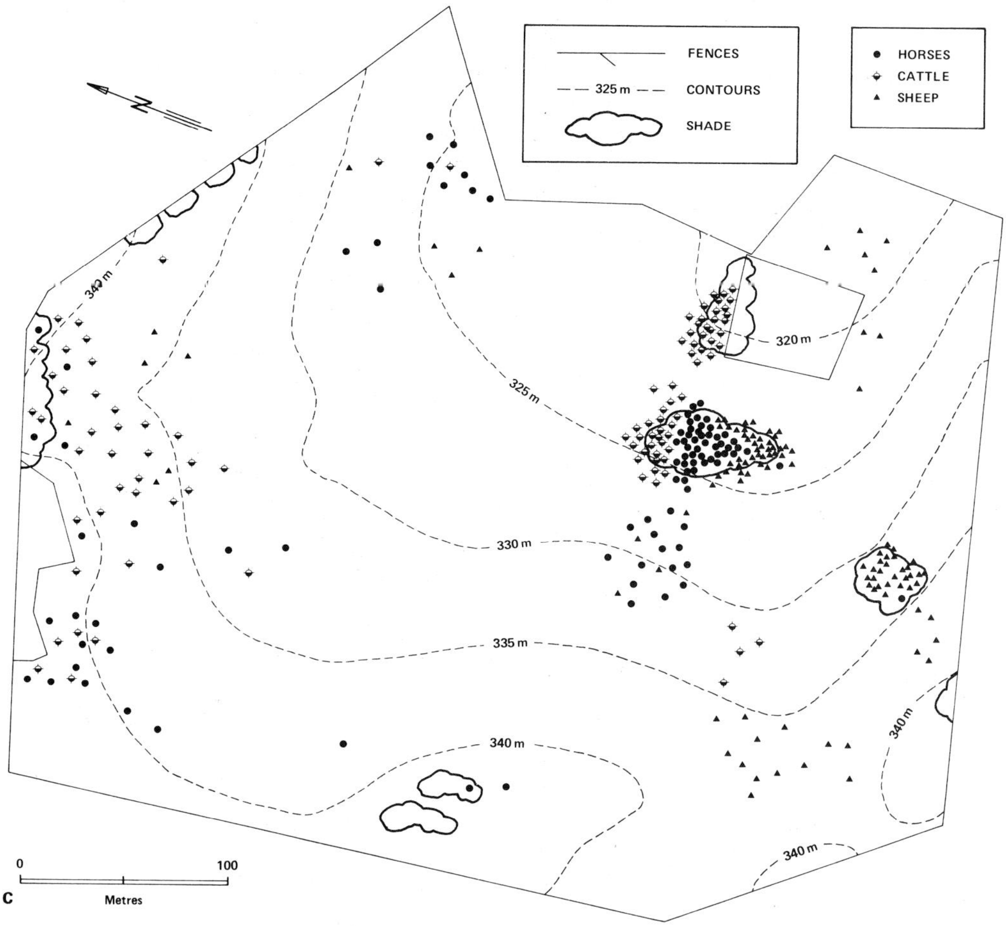

Fig. 1.22. (c) Resting during the day in summer.

Provision of shelter treebelts in winter may improve productivity of cattle as Bishop (1965) found on the treeless Eastern sourveld of South Africa. But in other areas artificial shelter has had no effect because cattle did not use the shelters even during winter storms, preferring natural shelter from vegetation and dunes (McIlvain and Shoop, 1971).

Sheltering behaviour of sheep grazing hill country in Britain has been studied intensively. Munro (1962) studied Scottish Blackface ewes on a 15 ha paddock in the Eastern Highlands of Scotland. Dips and hollows in the ground, and rocks were the only sort of shelter available. Each sheep's location was studied at 1000 h each day. When wind speed exceeded 38 km per hour the sheep were found to be on sites that were >50% less windy on 90% of occasions, but at lower wind speeds there was no relationship between

Plate 1.2. Sheep, cattle and horses camping or resting in shade.

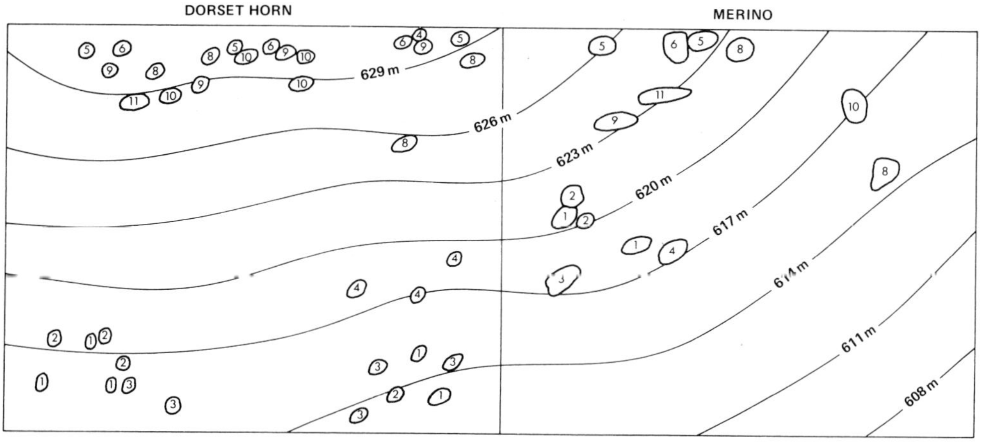

Fig. 1.23. Changes in location of camping sites of sheep in winter with changing temperature. *Note*: Each number represents a different week. The mean grass temperatures in those weeks were

1	2	3	4	5	6	7	8	9	10	11
−4.6	−2.8	−3.4	−10.9	−10.0	−6.5	−5.9	−9.3	−5.2	−5.9	−4.5

Plate 1.3. Sheep camping on the bare ground of a dam wall.

Plate 1.4. Sheep digging camping holes.

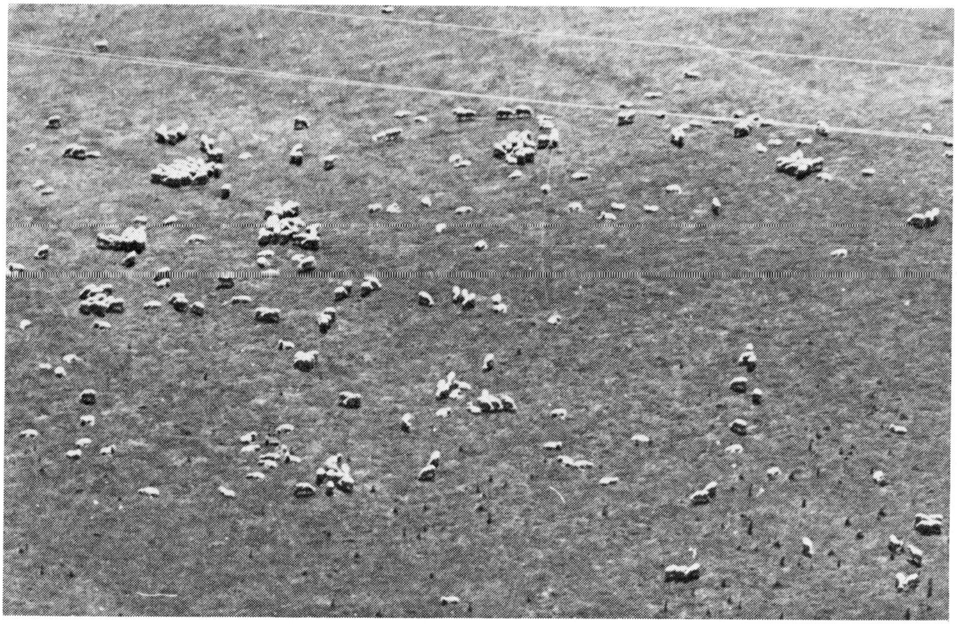

Plate 1.5. Shorn sheep clustering together with their heads under each other's bellies.

wind speed and the degree of shelter the sheep had sought. A distinct preference was shown for grassy hollows, as opposed to rocks. Temperature affected sheltering from wind, the degree of shelter increasing when temperatures were $<0°C$.

Mountain sheep in Central Wales behave similarly (Griffiths, 1966). All sheep were found on the most sheltered area of the hill when wind speed exceeded 25 km per hour. However, the degree of exposure tolerated by sheep depends on the location of preferred grazing areas (Griffiths, 1967). Even if preferred grazing areas are exposed to the wind, sheep will continue to use them except in severe conditions.

Welsh mountain ewes do not seek shelter when lambing when the wind speed is <8 km per hour or the Kata cooling power is below 20 millical cm^{-2} sec^{-1} (Winfield et al., 1969). As these values are progressively exceeded, lambing sites chosen are progressively more sheltered. However, individual ewes vary greatly in the degree of shelter they seek.

In a milder climate in the south-west of Western Australia, Arnold and Charlick (unpublished) found that Merino ewes having a choice of lambing in the shelter of trees or in the open showed no preference for either when average temperatures ranged from 8 to 15°C; only when subject to cold stress soon after shearing did they prefer the sheltered areas. Under colder conditions on the New England Tablelands of New South Wales, Lynch and

Alexander (1976) obtained similar results when shelter was available from tall dry grass.

Artificial shelters were not used by Clun ewes in winter (Powell, 1966) although their lambs made use of them. The lambs appeared to be most sensitive to cold in the first 14 days of life. A similar situation was found with Scottish Blackface ewes in spring in Northumberland (Miller, 1968). Weather conditions affected the use of shelter by lambs, with 24% sheltering when it was raining, but only 6% in cloudy or sunny weather. When it was raining the proportion of lambs sheltering increased at higher wind speeds (26—43 km per hour).

The provision of artificial shelter has long been a tradition in the hill country of the Scottish Borders where most hills have a stell — a circular stone shelter (Plate 1.6). The siting of these is of fundamental importance. The most valuable stells are sited on the lower slopes of hills, not in the valleys (Munro, 1961). They have to be high enough above the valley floors to provide access to grazing blown clear of snow. They provide shelter from wind from all directions.

Shelter from cold wind needs to be near to the edge of a paddock furthest from the direction of the wind. This is because both cattle and sheep move

Plate 1.6. A stell used by sheep for shelter. (Courtesy Hill Farming Research Organisation, Edinburgh, Great Britain.)

Plate 1.7. Sheep huddling together in the corner of a paddock, driven there by wind and rain.

with the wind under such conditions. A lack of shelter in this part of the paddock results in animals huddling together in a corner (Plate 1.7) if the wet, windy conditions persist for any length of time.

WALKING AND DRINKING

These two aspects are very often associated when food is in short supply. On extensive rangeland, the animals may have to compromise between the preferred frequency of drinking and the distance to travel from water to reach least heavily grazed areas. The interaction is of very considerable significance in rangeland management and in conservation of soil and vegetation. It is discussed in detail in Chapter 6.

In small paddocks with abundant green feed and water the distance walked by animals is small and involves mainly movement whilst grazing. Under extensive conditions with natural water freely available the animals will range further for feed but need not travel far for water, but when only limited watering points and dry feed are available, walking to and from water accounts for most of the distance travelled.

TABLE 1.8

Recorded distances travelled by cattle in different environments

Location	Class of stock	Size of paddock (ha)	Distance walking	Source
Reading, Great Britain	Dairy cows	0.1	0.9	Castle et al., 1950
Ayr, Great Britain	Ayrshire	0.5—0.8	1.2—2.0	Waite et al., 1951
Hurley, Great Britain	Hereford steers	0.2	2.1—2.7	Hughes and Reid, 1952
Alaska, U.S.A.	Friesian heifers	∞	2.6—6.9	Compton and Brundage, 1971
Colorado, U.S.A.	Heifers	20	2.4 Low SR * 3.2 Med SR 3.2 High SR	Quinn and Harvey, 1970
Montana, U.S.A.	Hereford cows	41	3.2	Peterson and Woolfolk, 1955
North Dakota, U.S.A.	Hereford cows	12 41 260	2.7 5.0 8.9	Shepperd, 1921
California, U.S.A.	Hereford cows	94—203	1.3—3.5	Wagnon, 1963
Kansas, U.S.A.	Hereford cows	338	4.0—4.8	Moorefield and Hopkins, 1951
Oklahoma, U.S.A.	Hereford cows	610	5.0	Dwyer, 1961
Texas, U.S.A.	Hereford cows	691	6.6	Box et al., 1965
Utah, U.S.A.	Cows of several breeds	700	3.9	Gonzales, 1964
Oregon, U.S.A.	Hereford steers Brahman × Hereford steers	813	7.4 9.7	Sneva, 1970
New Mexico, U.S.A.	Hereford cows Santa Gertrudis cows	1464	7.9 12.6	Herbel and Nelson, 1966
Transvaal, S. Africa	Ayrshire cows	—	1.8	Bonsma and Le Roux, 1953
Northern Territory, Australia	Shorthorn cows	∞	3—9.2	Schmidt, 1969
Utah, U.S.A.	Hereford	2000	6.3	Malechek and Smith, 1976

* SR = Stocking rate.

Distance walked

A survey of the recorded distances walked by cattle and sheep is shown in Tables 1.8 and 1.9. The variation is due to combinations of the factors mentioned above. The values do not represent the potential distances that animals can walk: for example, cattle have been recorded travelling 26 km a day under drought conditions on rangeland (Bonsma and Le Roux, 1953) although it is unlikely that such a distance would be travelled every day because drinking frequency is less than daily in such circumstances (Schmidt, 1969). Experimental studies have shown that Border Leicester sheep will travel 14 km, Merinos 13.7 km and Dorset Horns 9 km a day (Squires et al., 1972). In the Algerian Sahara camels rarely graze more than 30 km for water in summer, but in the Western Sahara they may move in excess of 80 km (Gauthier-Pilters, 1974).

When grazed on high yielding temperate pastures in small paddocks, dairy cattle walk up to 2.7 km a day, but under extensive conditions this increases to up to 9 km. The effect of paddock size was demonstrated by Shepperd (1921) who found that on a 12 ha paddock cattle walked 2.7 km and on a 260 ha paddock 8.9 km. The latter value is close to the maximum average for unfenced areas of unlimited size as studied by Schmidt (1969) where the limitation is imposed by the distance cattle will regularly travel for food

TABLE 1.9

Recorded distances travelled daily by sheep in different environments

Location	Size of paddock (ha)	Distance walked (km)	Source
Stratford, Great Britain	0.2	1.5—2.7 Winter 0.7—1.6 Summer	Hughes and Reid, 1952
Aberdeen, Great Britain	0.4	3.3—5.3	Tribe, 1949
Palmerston North, New Zealand	0.4—20	0.4—3.1	Cresswell, 1960
Groot Fontein, S. Africa	125	5.2—8.0	Louw et al., 1948
Oregon, U.S.A.	1000	1.6	Jardine, 1910
Utah, U.S.A.	1134	2.9—5.1	Bowns, 1971
Montana, U.S.A.	1620	7.6	Van Dyne and Van Horn, 1965
Ivanhoe, Australia	2000	3.2—5.5	Lynch, 1974
Deniliquin, Australia	1310 910	8 —14 4.7—6.2	Squires, 1976

from a single watering point. Shorthorn cattle grazing in a 15 000 ha paddock in Central Australia were found to walk between 5 and 14 km a day (Low et al., 1978a).

There are differences in the distances different breeds of cattle walk under similar extensive conditions. Brahman × Hereford cattle walked 2.3 km further each day than Hereford cattle in one study (Sneva, 1970) and in another (Herbel and Nelson, 1966) Santa Gertrudis cattle walked 4.7 km more than Hereford cows. This is the only quantitative evidence that exists but it is generally accepted that the Brahman, Santa Gertrudis and Africander cattle will range farther than British breeds of cattle in hot climates. One reason is their greater heat tolerance, which was demonstrated earlier in this chapter on their grazing and resting habits.

Within a herd different animals may travel substantially different distances. A detailed study of Shorthorn cattle was made on the Barkly Tablelands in the Northern Territory of Australia (Schmidt, 1969). Schmidt found that in very hot conditions, and on areas with no shade, the herd split into two nearly equal sized groups. One of these, the 'non-walkers' grazed close to the watering point, travelling a maximum distance of 4 km from water, and 9.6 km a day, grazing on sparse feed. The remainder of the herd, the 'walkers' averaged 16 km a day: they began walking along paths and grazed for much of the night in a small area. Sometimes this group would regather after a period of grazing and walk a further distance before grazing again. The only discernible difference between the groups was the proportion of cows without calves, this being 19% in the 'non-walkers' and 39% in the 'walkers'.

The distance walked by sheep on rangelands depends on the type of vegetation and frequency of drinking. On a saltbush paddock Merino sheep in summer walk 8 km a day when they drink once a day and 14 km a day when they drink twice a day; on a grassland paddock drinking is once a day and the sheep walk 4.7—6.2 km a day (Squires, 1976). Several studies have shown that breeds of sheep differ in the distance walked each day. Differences in the potential of Border Leicester, Merino and Dorset Horns have been mentioned already. In South Africa Persian Blackhead sheep or their crosses with Dorset Horns walked up to 2.8 km farther in a 125 ha paddock than Karakul or Merino sheep (Louw et al., 1948). On a mountain range in Utah, Columbia, sheep averaged 3.0 km a day, Targhees 3.8 km and Rambouillet 4.6 km over a grazing season (Bowns, 1971). These differences in walking resulted in the Columbia sheep having a camping site lower down the mountain slope than sheep of the other two breeds.

In New Zealand, Romney and Cheviot sheep were compared on both flat and hill country (Cresswell, 1960) and an interesting difference was seen in the amount of walking done. On hilly country the Cheviots walked 57% further than the Romneys, but only 21% further on flat country. Sheep of both breeds walked further on flat country than on hilly country.

There are few reliable estimates of rate of walking. For cattle, Herbel and

Nelson (1966) and Box et al. (1965) give distances travelled and time spent travelling which suggest speeds between 4 and 7 km per hour. Squires (1974) has made a detailed study with Merino sheep. He found that as sheep grazed towards water they gradually increased the walking component. Thus their speed was 0.6 km per hour when 29—15 min from water and then increased to 4.3 km per hour as they moved in to drink. Returning from water to a grazing area they walked at about 2.5 km per hour.

There has been only one study comparing different animal species: Cory (1927) found that on rangeland in Texas, Hereford cattle averaged 5.3 km, Rambouillet sheep 6.1 km and Angora goats 9.6 km each day. The only other figure seen for goats on rangeland is 4.5 km (Askins and Turner, 1972).

In none of these studies have reasons been found for the differences, but length of step may be important. For example, Border Leicester sheep have much longer legs than Dorset Horn sheep and can presumably take longer steps. On the other hand, general conformation and musculature could be important.

On rangeland the time taken by stock in walking to water can be considerable. Herbel and Nelson (1966) found that Santa Gertrudis cattle spent on average 2.9 h a day walking and Squires' (1974) data indicate that Merino sheep walking 7 km a day may take 3 h doing so. In later work (Squires, 1976) on two types of rangeland, up to 6.7 h were recorded in walking. Bowns (1971) gives figures of 1 to 1.5 h a day for sheep on rangeland in Utah.

Frequency of drinking and quantity of water drunk

The frequency of drinking will depend on temperature, condition of feed, and the distribution of water. When grazing on abundant green feed sheep and cattle may not drink water at all, whilst on dry feed they will need to drink regularly. Drinking frequency increases with temperature and is more frequent in small paddocks or where there are several watering points. The more water is available, the more drinking will occur. A high salt content in the feed, such as occurs in *Atriplex* shrublands, will cause more frequent drinking.

The time of day when stock drink must be related to their daily patterns of grazing and resting, but there is relatively little quantitative information on either the frequency of drinking or on when it is done.

Dairy cows grazing ryegrass pastures in Scotland were found to drink 24 litres of water a day when they were yielding 15 litres of milk (MacLusky, 1959). The cows averaged four drinks a day, but individuals varied from one to six drinks. The only other study under intensive grazing was with Charolais cows with calves grazing Kentucky Bluegrass pasture in Pennsylvania (Gary et al., 1970); they averaged 3.5 drinks a day. Beef cattle grazing rangeland in the U.S.A. in summer drink between one and three times a day (Dwyer, 1961; Gonzales, 1964; Herbel and Nelson, 1966; Sneva, 1970) and in winter only twice (Box et al., 1965). Gonzales (1964) noted a seasonal variation in his studies in Utah but no clear pattern from year to year. The range in his study was 1.3—2.7 in one year with an average of 1.9, and 1.6—2.8 in a second year with an average of 2.3.

Shorthorn cattle under hot dry conditions in Northern Australia come in to water at least once daily unless they have travelled an unusually long way from the water supply; in this case, they come in only every second day (Schmidt, 1969). Once at the water the animals may drink several times over a period of several hours. An observation on 18 cattle on one day showed that four drank four times, seven drank three times, six drank twice and one had only a single drink. Low et al. (1978a) made a more detailed study in Central Australia (Table 1.10) which showed that in summer 84% of animals watered once a day or once every two days, whereas in winter 44% watered only every third or fourth day.

The pattern of watering with cattle varies little on rangeland, with the first drink occurring after the major morning period of grazing. Dwyer (1961) noted that the cows made visits to water between 0700 and 0800 h but did not remain there, but on their return from grazing at 1100 h they remained resting near the water for several hours. Sneva (1970) found that 30% of drinking was done in summer between 0600 and 1200 h, 53% between 1200

TABLE 1.10

Frequency that individually tagged cattle were seen drinking during four-day observation periods on Kunoth Paddock, 1970—1972 (Source: Low et al., 1978a)

	Vegetation [1] condition index	Average [2] maximum daily temperature	Frequency				*n*
			% every 1st day	% every 2nd day	% every 3rd day	% every 4th day	
Summer	1	36	80	11	7	1	141
	2	33	33	35	32	—	137
	2	35	49	44	7	—	68
Winter	2	21	37	27	19	17	59
	4	17	16	32	30	21	164
	4	24	33	22	30	16	212
Spring	3	33	12	19	39	30	242
	4	35	42	33	17	8	112

[1] The vegetation condition index was rated on the basis of phenological state and abundance of forage, 1 being most preferred, 4 the least.

[2] Average maximum daily temperature determined from maximum temperature on day preceding watch plus four days of watch.

and 1600 h and 17% between 1600 and 2000 h. Weaver and Tomanek (1951) found that the cattle they studied in Nebraska camped near water most of the day. This was the pattern also found by Schmidt (1969) under hotter conditions when cattle had no shade. Where there was shade, cattle became either day waterers or night waterers, all animals showing a clear preference for one or the other. Fig. 1.24 shows the diurnal pattern of cattle coming in to water and of drinking on one day on one of Schmidt's study areas. The main drinking period was between 0600 and 0800 h. Fig. 1.25 shows how the diurnal pattern of watering in a herd changes from season to season in Central Australia, the bimodal peaks being further apart in summer than in winter.

Sheep appear to drink with similar frequency to cattle. Merino sheep grazing abundant feed on a *Casuarina—Heterodendron* association in summer were found to visit the watering point only three times in 11 days, and in winter did not drink at all in an 11 day period (Lynch, 1974b). Presumably the moisture content of the food was high and it supplied most of the animals' water requirements, even in hot conditions. By contrast, sheep grazing *Atriplex* shrublands have been seen to drink twice a day (Squires, 1976) depending on the distance between food and water (Squires and Wilson, 1970).

Bowns (1971) found that 73% of drinking by sheep on summer range in Utah was done before 0830 h. A similar pattern was found in Northwestern Australia by Arnold (unpublished) with 64% occurring by 0800 h, although a further period of drinking occurred between 1430 and 1700 h. Squires (1971) found that sheep drinking twice a day did so at around 0800 and 1600 h in summer. Merino sheep drinking twice a day when the maximum temperatures exceed 38° C abandon their afternoon walk to water (Daws and Squires, 1974). Border Leicester sheep responded to higher temperature by drinking earlier in the morning.

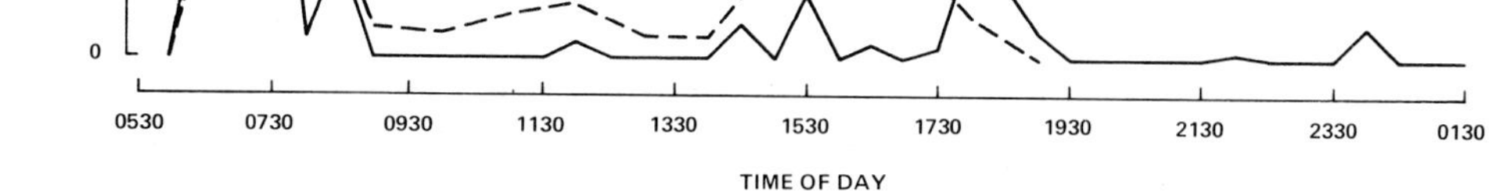

Fig. 1.24. Diurnal pattern of arrival of cattle at water and of drinking — Barkly Tablelands, Northern Territory, Australia. (Source: Schmidt, 1969.)

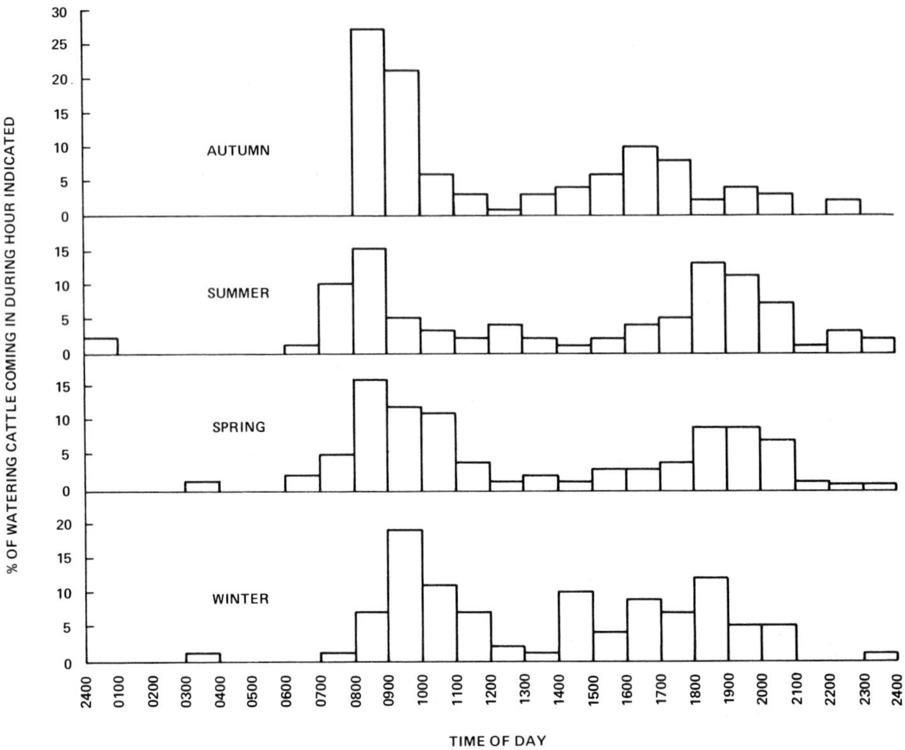

Fig. 1.25. Diurnal pattern of drinking by cattle in different seasons at Alice Springs, Australia. (Source: Low et al., 1978a.)

Under more intensive conditions Arnold (unpublished) has compared Merino and Dorset Horn sheep grazing on 18 ha paddocks in Western Australia. These sheep did not drink during the period when pastures were most succulent, but for the drier 10 months of the year the number of drinks per day varied from 0.3 to 6.0. Dorset Horns average 2.4 drinks a day and Merinos 3.4. However, the number of drinks increased as maximum temperature increased and decreased with increased humidity (Fig. 1.26). The two breeds had different diurnal patterns of drinking (Fig. 1.27) with a less distinct diurnal pattern in Dorset Horns; Merinos drank between 1800 and 0500 h whereas some Dorset Horn sheep were drinking during every hour of the day and night.

Wilson (1974) found that on *Danthonia caespitosa* grassland, water was consumed for up to 7 months each year. Maximum intakes were 3—3.5 litres per sheep per day, but this was reduced for 1—2 weeks after falls of rain of less than 25 mm and for longer periods after heavier falls.

On the saltbush (*Atriplex vesicaria*) and belah—rosewood (*Casuarina cristata—Heterodendrum oleifolium*) communities, water intakes were up to

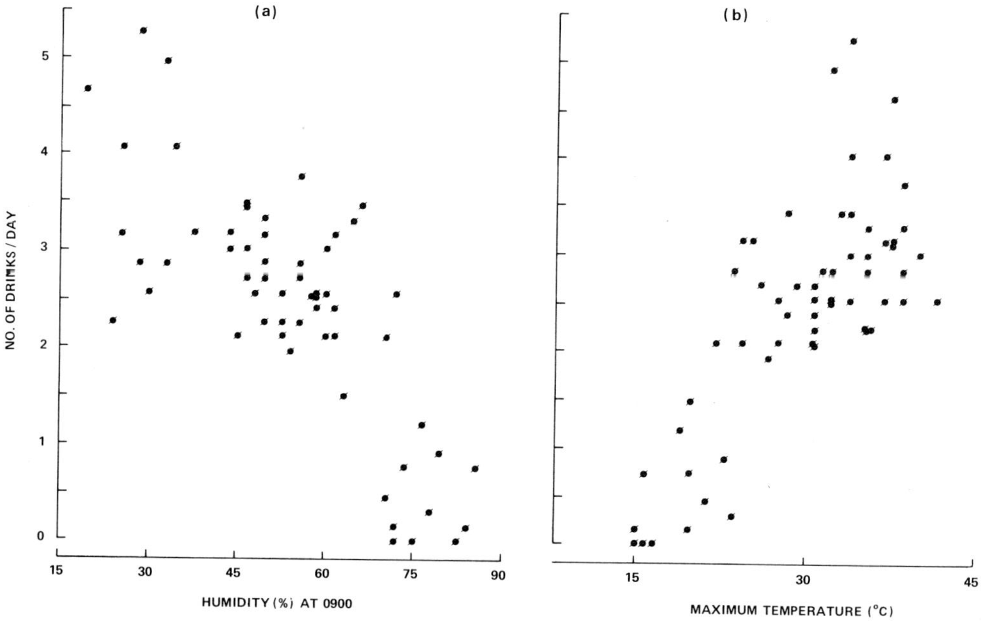

Fig. 1.26. Changes in frequency of drinking in Merino and Dorset Horn sheep with (a) humidity and (b) temperature.

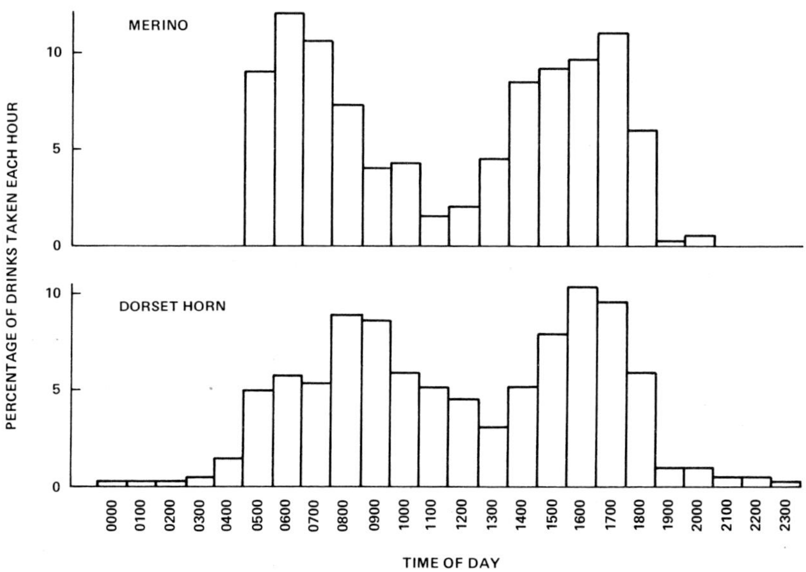

Fig. 1.27. Diurnal pattern of drinking in summer of Merino and Dorset Horn sheep at Baker's Hill, Western Australia.

6—7 litres per day in summer. In one particularly dry summer water intakes of the sheep on saltbush rose to 12 litres per day for several months. The higher water intakes of these sheep were attributed mainly to the high mineral content of the *Atriplex* and *Bassia* spp. eaten by them, although the sheep on the belah—rosewood community were also in a hot environment. In contrast to the sheep on *Danthonia*, those on the bush communities consumed significant amounts of water throughout the winter in years of low rainfall.

Camels can exist for months without drinking even when grazing on dry mesophyte pastures, the herbage providing 3—30 litres a day. During the cool season when herbage moisture content is 40—60% they may go up to 6 months without drinking (Gauthier-Pilters, 1974). When the temperature rises to 30—35° C they start visiting drinking places, but only when day temperatures reach 40° C and night temperatures 30—35° C does their rhythm become regular, i.e. once every 3—4 days. When they drink they may consume a third of their bodyweight in 10 min.

ELIMINATION

There are no reports of defaecation frequency in free-ranging horses. In confinement Dukes (1955) recorded 5—12 defaecations and Bulbenik (1961) six for mares; dominant stallions averaged 20 because they covered each defaecation by the mares in their harem.

The number of defaecations in 24 h has been recorded for cattle but not for sheep. The range in mean frequency reported in 20 reports is 2—16 times a day. Most reports give an average of between 11 and 13 times a day. Several factors influence frequency of defaecation in cattle. Defaecation is more frequent when cattle are grazing green than dry forage. Wagnon (1963) reports frequencies of 9—13 times a day on green forage and four on dry forage. Moisture content of the food consumed would be higher on green forage, not all of it being eliminated in urine. Stocking rate has an effect when animals consume less feed at a high stocking rate. Thus Wagnon (1963) found that cows defaecated nine times a day at a high stocking rate and 12 times a day at a low stocking rate.

The diurnal pattern of defaecation is not clear. Dwyer (1961) observed a greater frequency when cattle were grazing (eight times in 9.7 h) compared with four times during a resting period of 13.4 h. On the other hand, Gonzales (1964) reported that defaecation was most common at resting areas. However, he gave no figures to substantiate this statement. In New Zealand, Hancock (1950) found that 46% of defaecations occurred in the 16 h between 15.00 and 07.00 h, which supports Dwyer's finding because grazing at night is less than during the day. Suzuki et al. (1967) made a similar observation.

Frequency of urination is lower than that of defaecation, the range reported being 2—12 times a day, with only three reports of values greater than ten. Cows grazing rangeland in California urinated less when on dry forage than when on green forage, and less when at high than at low stocking rate (Wagnon, 1963). The diurnal pattern has been examined only once.

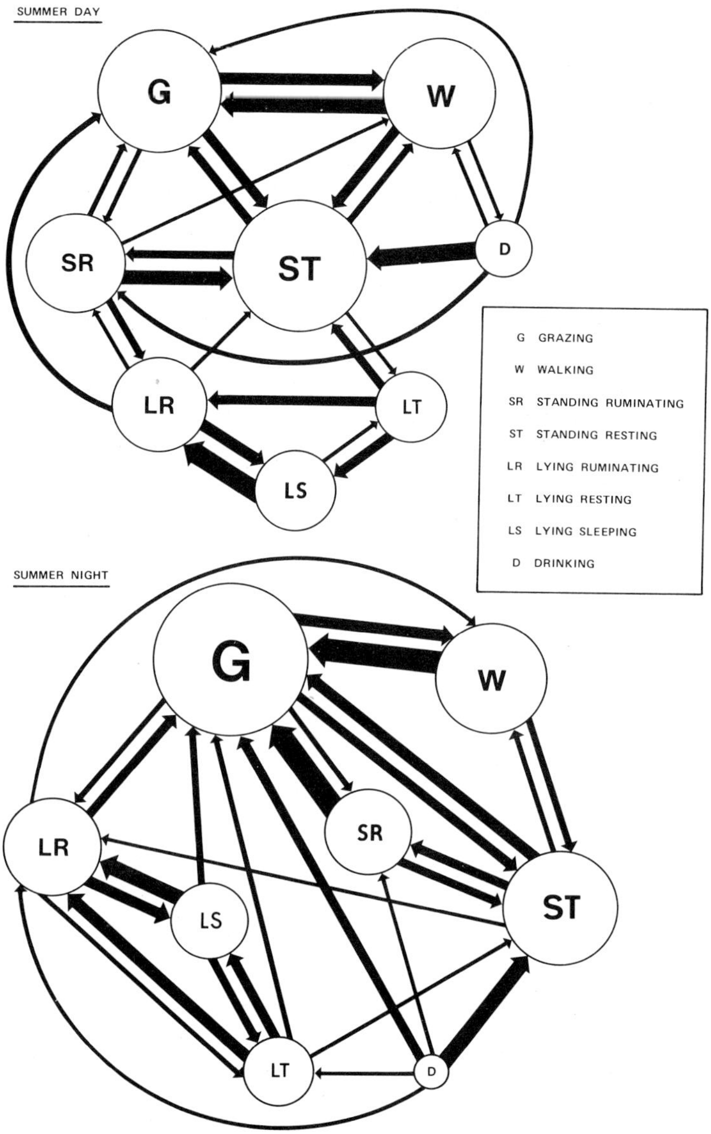

Fig. 1.28. Ethogram of transition probabilities between activities for Shorthorn cows in Central Australia in summer, (Source: Low et al., 1978b.) The size of the circles indicates the number of times a given behaviour trait was a transition to another trait.

Dwyer (1961) recorded that only a third of urinations took place when cattle were resting. In horses Dukes (1955) recorded 7--11 urinations and Bulbenik (1961) five for mares and 27 for the dominant stallion which covered the mares' urination areas. The reason for this and the covering of faeces is not clear since territorialty is not involved. It may be that it is done to keep other stallions away from the stallion's harem.

INTERRELATIONSHIPS

We have shown that the various activities such as grazing, ruminating, lying, standing and drinking vary in duration and timing depending on environmental conditions. Can the pattern of activities illustrated in Fig. 1.1 be predicted? Recent work (Low et al., 1978b) with Shorthorn cattle in Central Australia suggests that the patterns do have some consistency. Using sophisticated statistical techniques they obtained the probabilities of one activity occurring after another activity as shown in the ethograms in Fig. 1.28. During the night grazing assumes a 'central' position in the activity pattern, that is, most other activities lead with a fairly high probability to grazing. During the day standing assumes this 'central' position which might be called a 'transitional' activity.

Chapter 2

SOCIAL ORGANISATION AND ANIMAL DISPERSION

Animals are not dispersed randomly in any environment, and free-ranging domestic animals may exhibit extreme non-randomness in the use of the resources of the environment, particularly the vegetation, with large effects on the environment that are discussed in Chapter 6. This non-randomness in use is related to the physical and biological structure of the environment, to weather and to patterns of maintenance and social behaviour.

In this chapter the basic patterns of social organisation, based on relationships between individuals, and including the effects of man's husbandry practices, are considered first. Factors affecting dispersion of flocks and herds in different environments are then discussed. Finally the interactions between distribution of different types of vegetation and animal dispersion are considered in relation to the use of the vegetation.

BASIC PATTERNS OF SOCIAL ORGANISATION

Social structures

In feral flocks of sheep like the Soays of St. Kilda, goats, semi-wild cattle like those in the Camargue of France, and wild horses the individuals form up into home range groups (Riney and Caughley, 1959; Schloeth, 1961; Grubb and Jewell, 1966; Hall, 1972). In sheep and goats the females and the juveniles use certain fixed areas, which may change with season, with the males associating in bachelor groups in other areas but joining the females during the rut. In these groups the basic social unit is the female with its most recent offspring; more loosely associated with this unit but within the home range area, are likely to be related animals, for example a yearling and the female's mother. Thus the home range group is usually made up from numbers of sub-groups of related individuals. In the bachelor male groups the associations between individuals are looser and the composition of the groups may change frequently.

In feral horses, harems are formed, with younger males in bachelor groups; the males in all species either being forced out or voluntarily leaving female groups before reaching sexual maturity. Some older male horses may live solitary existences.

In husbanded flocks and herds the males are either castrated or separate except for a period of mating, and the juveniles are 'weaned' and put together to form a flock or herd. There are variations on this; for example in the hills of Britain the juvenile females remain with their dams and only the males are removed.

The social organisation of a flock or herd depends basically on recognition by the individual of its status in relation to other individuals. It has been shown that a group of sheep of a single breed put together into a paddock develops a flock or group identity so that when other sheep are put into the paddock, even if they are of the same sex and age, it takes several weeks for the two groups to become integrated (Arnold and Pahl, 1974). This is illustrated in Fig. 2.1. It has been shown that this group identity is due in part to a group odour because anosmic sheep mix freely with strange sheep. Sheep recognise and prefer to associate with sheep of the same breed when given the choice (Arnold and Pahl, 1974). In one study it was shown that ewes of seven breeds, which had grazed together as a flock for a year still showed a significant preference for members of their own breed.

If the paddock into which sheep of different breeds or histories are put is large then each breed may use different parts of the area, and they may

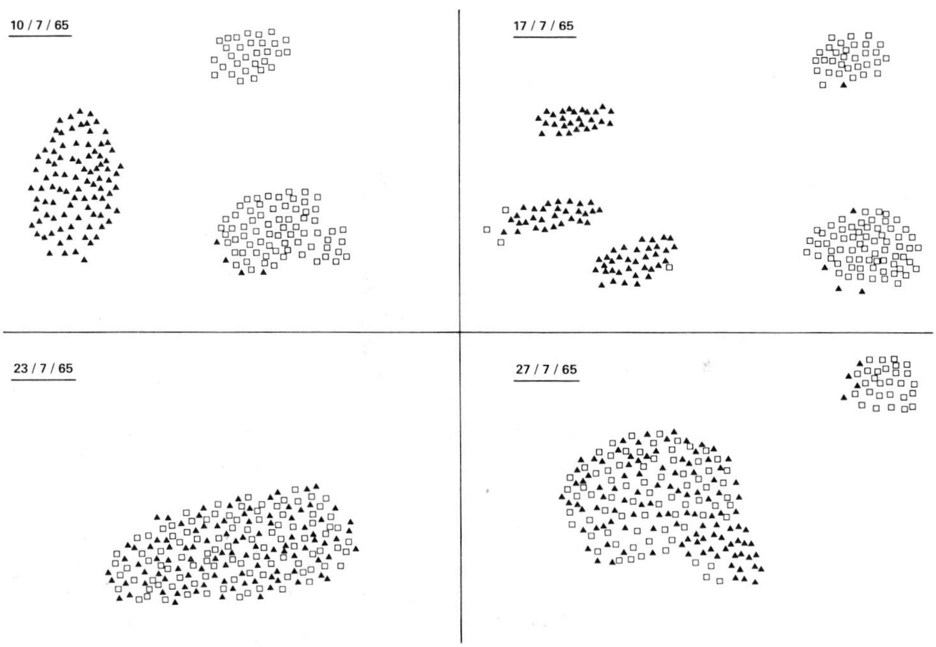

Fig. 2.1. Stages in the integration of two groups of Border Leicester and Merino sheep to form a flock. (Source: Arnold and Pahl, 1974.) *Note*: The diagrams are four of the maps made when sheep were camping. The sheep were put together on 10/7/1965. The ▲ and □ denote the two groups of sheep.

never mix. Hunter (1960) observed that North Country and South Country Cheviots grazed in the same paddock for four years and never mixed.

The discussion so far shows that some social organisation exists in husbanded flocks. It is possible to trace how this social organisation develops. The first social bond a sheep develops is with its dam, and this is described in detail in Chapter 5. Once the bond is established it will usually remain in the females unless broken by separation, though the behavioural relationships change with time. Merino ewes and lambs during the first 4 weeks of the lamb's life are found to stay within 10 m of each other for over half the time, no matter what the activities of the two animals (Morgan and Arnold, 1974). The distance at any one time of the lamb from its dam depends on the activity of the ewe; if the ewe is lying or walking then the lamb is usually within a metre. If the ewe is grazing then the lamb has greater freedom of distance and activity (Table 2.1). The distance between ewe and lamb when both are grazing increases rapidly over the first ten days of life, reaching a plateau with an average distance of 20 m (Fig. 2.2). This then is the average social distance between dam and offspring for Merinos when grazing on pastures of abundant feed.

During the first weeks of life lambs, calves and foals form peer groups. In Merino lambs it was found that at 4 weeks of age 80% of lying, 63% of grazing and 100% of playing was done in peer groups (Morgan and Arnold, 1974). We found that from 6 to 12 weeks of age Merino lambs kept nearer to each other when grazing (average 7 m) than to their dams (20 m). But the lambs retain contact with their dams, their dam usually being the nearest ewe. At a later stage the association with peers seems to become less important although there is little quantitative evidence on this aspect because the lambs are often separated from their dams at this age. It is from this mother—offspring relationship that the matrilinear groupings form in home range groups when weaning by separation does not occur.

The formation of 'weaner' flocks breaks the dominant social bond in the young animal's life and a new social organisation has to be developed. This

TABLE 2.1

Relationship between ewe and lamb behaviour. Numbers of observations of lamb behaviour for each ewe behaviour trait (Source: Morgan and Arnold, 1974)

Ewe behaviour	Lamb behaviour					
	Lying	Standing	Grazing	Walking	Sucking	Playing
Lying	693	37	7	3		3
Standing	351	264	28	8	417	16
Grazing	1978	549	629	70		228
Walking	22	10	1			4

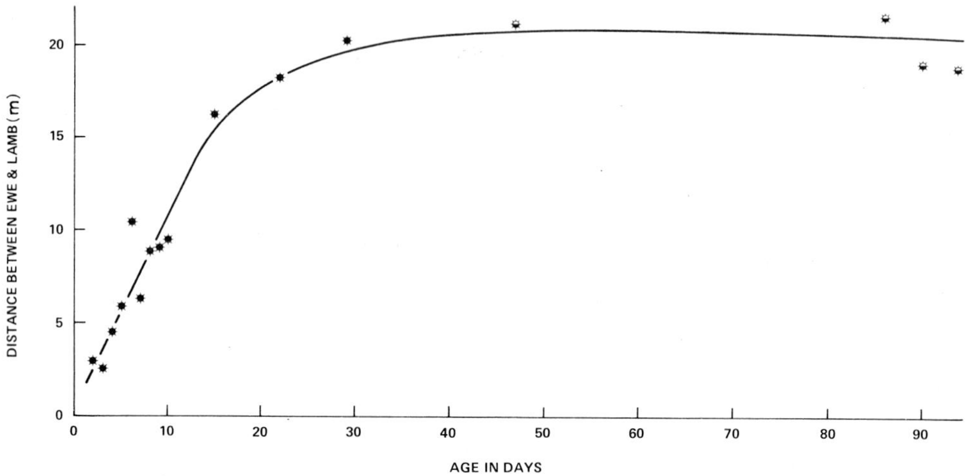

Fig. 2.2. Changes with lamb age in the social distance between Merino ewes and their lambs when grazing. Source: ✸ Morgan and Arnold, 1974; ✸ Arnold, unpublished.

invariably begins with the formation of small groups in which inter-animal distances are low. Gradually, these groups become larger until a flock is formed. It is clear to the shepherd that weaners take time to develop a flock identity since when mobbed up they form into not one but several aggregates, unlike adult sheep where, in Merinos, for instance, any groups within contact distance immediately run together.

The size of sub-groups increases with age from separation at 4 months (Table 2.2). This is unrelated to the size of the paddock or space available to the animals, but even as late as 11 months of age sub-groups may be formed. Normal adult flocking behaviour appears to be established by 15 months of age. In some breeds of sheep sub-grouping always occurs and this seems to

TABLE 2.2

Mean group size for flocks of Merino sheep with increasing age at two different locations (Source: Arnold and Pahl, 1967)

Age (months)	Canberra (16 ha)	Deniliquin (293 ha)
4	8.4	17.4
5	—	6.2
7	18.0	14.6
8	25.3	—
9	—	25.2
11	34.0	36.0

TABLE 2.3

Mean group sizes for flocks of 80 eight-month-old sheep of different breeds (Source: Arnold and Pahl, 1967)

Flock	Mean group size
Merino	32.8
Border Leicester × Merino	29.4
Dorset Horn × Border Leicester × Merino	16.3
Dorset Horn × Merino	13.9

be an inherent characteristic that is seen from an early age (Table 2.3).

It is necessary to consider first the basic pattern of flocking behaviour and inter-animal distances firstly under abundant feed conditions because environmental factors cause major changes in flocking and herding behaviour: changes which will be discussed later.

There are three characteristic flock structures: a tightly knit flock, typified by Merino sheep (Plate 2.1); a flock widely dispersed, but with uniform spacing between individuals, such as Border Leicester sheep form, and a flock split up into sub-groups as seen in Swaledales (Plate 2.2) and Dorset

Plate 2.1. Dispersion in a flock of Merino sheep grazing abundant pasture.

Plate 2.2. Dispersion in a flock of Swaledale ewes.

Horns. Where sub-grouping occurs the flock remains a social entity because the membership of the sub-groups is constantly changing.

Within these flocks of young sheep, individuals may show preferences for grazing within groups of a particular size (Arnold and Pahl, 1967). However, in only a few breeds do individuals form bonds with other sheep so that pairs of sheep are found together consistently (Arnold and Pahl, 1967; 1974). In a flock of year old sheep containing 11 sets of twins it was found that in only three cases was the highest frequency of association between pairs of individuals between twins. This is interesting because it might have been expected that during early life a twin would form a closer bond with its twin than with other sheep of its own age, and that this bond would be strengthened when the young sheep were separated from their dams. In adult flocks of Dorset Horn and Merino sheep around 50% of individuals form close associations when grazing: it is not known whether the same associations occur when sheep camp. Associations between two or more individuals are maintained for limited periods of 1—2 months and then new associations are formed (Arnold, unpublished). Animals that are associating are found in the same groups in the case of the Dorset Horns, and in the Merinos, which always grazed as a single flock, the individuals associating grazed close to one another.

Identical twins in cattle associate closely together, especially when lying down (Petersen, 1956; Ewbank, 1967a). An individual appears to identify its

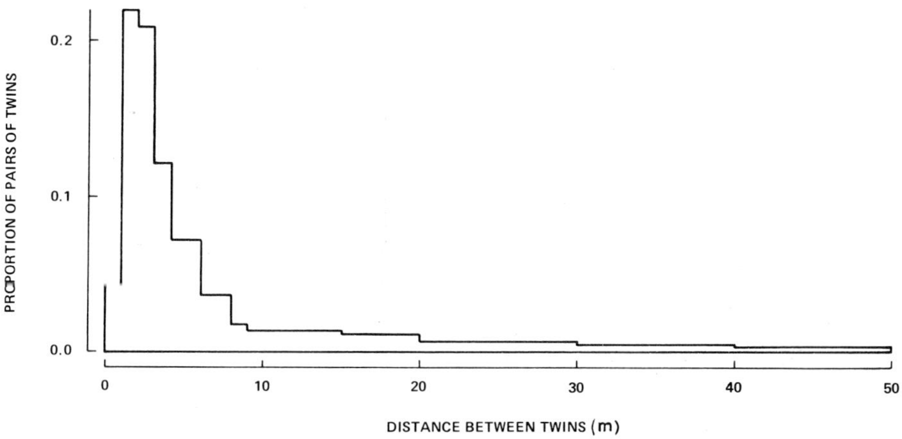

Fig. 2.3. The frequency of distances between identical twin cows when camped. (Source: Petersen, 1958.)

twin in the dark by its odour, and the pairs lie close together, usually within 2 m (Fig. 2.3). This association between twins in cattle probably arises because the pairs studied have been reared as pairs and do not know other individuals from birth, as occurred in the case of twin lambs.

In the Haute Savoie of France cows from many cow sheds are grouped and grazed as a single herd on alpine pastures in summer. The animals from a specific shed are to be found grazing together within the framework of the herd (Fayre, 1976).

In a herd of 15 mares and their foals, the foals, of 7—16 weeks of age, were found by Schoen et al. (1976) to associate in fairly fixed groups of two to three.

Mares that have been kept together will usually continue to associate consistently when put with other mares (Fig. 2.4). However, in a herd of mares made up from individuals from different studs, such close associations are not formed, although most individual mares are found to graze more often next to certain individuals. In herds of both sexes, colts and fillies tend to separate from the mares and stallion (Collery, 1974). The stallion usually attacks members of the younger group who approach too closely. Collery says that about ten animals seems to be the number of mares a stallion wishes to control. The stallion will round up the mares on the periphery of his herd or 'harem' but will ignore or ιepel fillies. In feral horses, Hall (1972) found the harems were of three to four mares, or if larger may contain two males.

In the absence of a stallion Collery observed a marked tendency for family groups to consort closely; a mare, her foal, yearling and two year old grazing near to each other. Once a mare foals the close association with her dam is usually allowed to lapse. A similar pattern occurs in the Exmoor ponies

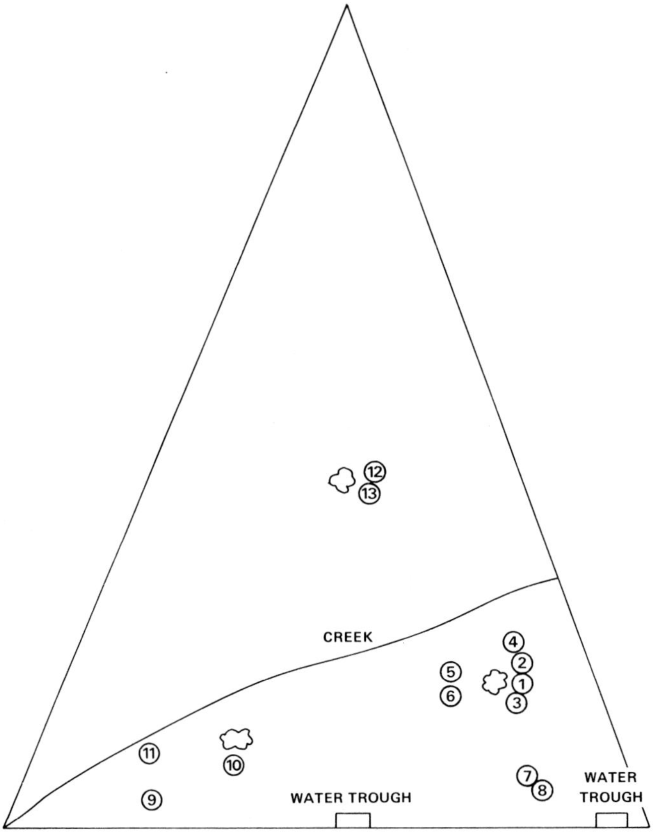

Fig. 2.4. Locations of individual horses when standing resting. *Note*: Prior to being put into the paddock the following horses had been associated with one another: 1, 2 and 4, 9 and 10, 12 and 13. Horses 7 and 8 entered the paddock on the same day but had no previous association. Horses 9, 10 and 11 had foals at foot but had no previous association.

(Tyler, 1972). Most groups are family groups (Plate 2.3), with fillies remaining with their mothers until their second, third or fourth year. Having left their mothers young mares frequently change groups, often ending up joining older mares that have foals or yearlings. The few stallions that are left on Exmoor in winter form bachelor groups, as occurs with feral horses on Sable Island, Canada (Bruemmer, 1967) and in the Pryor Mountains of Montana (Hall, 1972); the groups of young males have a much looser social organisation than that of the harems, with members leaving to form other groups for a period and then rejoining the previous group. In the forests of Tamasaki, Japan, the semi-wild mares are often found living as solitary individuals, but in the open ground they form groups (Imanishi, 1950). Males that have lost their harems often live as solitary individuals (Hall, 1972).

Plate 2.3. Family group of Exmoor ponies.

In the Sahara camels are allowed to free range for 4—5 months of the year. During this period they form mixed groups of females, juveniles and several males, led by an elderly female; small groups of females plus juveniles and bachelor male groups (Gauthier-Pilters, 1974). The groups are fairly fixed in composition, and if a camel is separated from a group for any reason, it will wait at a waterhole for the group to return, maybe a week later.

Social distance and communication between individuals

An important difference between breeds of sheep is in the distance between individuals in various activities. This has been measured in two ways, either by taking the average distance to the first nearest neighbour or by looking at the average area occupied by individuals within the flock 'boundary'. The average distance between neighbouring sheep when the animals are grazing varies from 4 m to more than 19 m (Table 2.4). The greatest distances are for the British hill breeds of sheep and the smallest for Merinos. The values given by Baskin (1975) are for herded sheep in Russia, which may not be spatially unconstrained.

The average distances do not convey the full picture. It can be seen from the distribution curves of the nearest neighbours in Fig. 2.5 that the majority

TABLE 2.4

Mean distances between nearest neighbours in sheep of different breeds

Country	Breed	Mean distance (m)
Great Britain	Exmoor Horn	8.6
	Herdwick	8.1
	Blackface	7.5
	Derbyshire Gritstone	7.5
	Kerry Hill	7.5
	Teeswater	7.2
	Rough Fell	7.0
	Welsh Mountain	6.9
	Wensleydale	5.7
	North Country Cheviot	5.5
	Swaledale	5.4
	Lowland Crossbreds	5.1
	Dalesbred	4.4
	Dartmoor	4.0
	Wiltshire Horn	3.4
	Suffolk	3.4
New Zealand	Romney	4.8
Australia	Merino	3.1
U.S.S.R. [1]	Karakul	1.7
	Fat tail	1.7

[1] Source: Baskin (1975).

of sheep in all breeds have a nearest neighbour within 5 m but that the breeds differ because of the proportions of sheep that allow a greater distance between themselves and the nearest other sheep. On this basis the breeds fall into four classes of dispersion. Merinos are the closest, lowland breeds are next, hill breeds like Swaledale, Rough Fell and Cheviot are further apart, with the mountain breeds (Blackface, Welsh Mountain and Herdwick) and Exmoor Horns being the furthest apart.

Distance to the nearest neighbour is only one attribute of dispersion. The cohesion of all members of a flock is another. This cohesion varies with environmental factors as will be discussed later, but the basic pattern can be assessed under ideal conditions — flat paddocks with abundant feed. Table 2.5 gives a set of figures obtained in small (1 ha) paddocks for the space occupied per sheep for nine breeds. The threefold difference between the breeds indicates large basic differences in dispersion between sheep of different breeds when grazing.

When resting, social distance is greatly reduced and in Merinos an analysis of 72 flocks showed that when camped the sheep occupied an area of 10 m^2 per sheep.

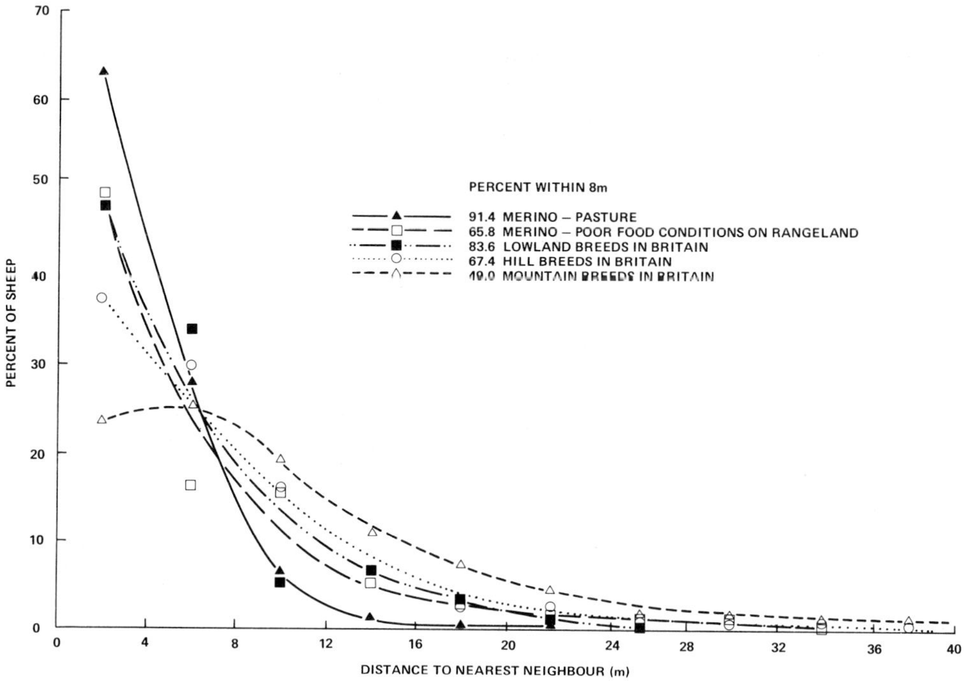

Fig. 2.5. Distributions of distances between nearest neighbours in different types of sheep.

Orientation of individuals one to another is probably important in maintaining social contact. Crofton (1958) observed that adult crossbred sheep in flat paddocks of 6—22 ha, were oriented when grazing, but not when engaged in other activities, in an arrangement whereby most sheep subtended

TABLE 2.5

Differences between breeds of sheep in dispersion

Breed	Average area occupied per sheep (m^2)
Cheviot	269
Border Leicester	340
Merino	410
Polwarth	416
Suffolk	466
Southdown	576
Corriedale	614
Romney	749
Border Leicester × Merino	883

two others, or one other and a prominent object such as a tree, with an angle of 111.9 ± 3.6°. Since 110° is the angle between the optical axes of the eyes, Crofton suggests that this position gives the sheep the easiest vision of its position relative to other sheep. The distances involved, between 14 and 27 m, are much greater than those obtained by the authors (Table 2.4). Weaner sheep did not conform, because, Crofton suggests, they bunch close together in small groups. It is an interesting theory but it is difficult to see the importance of the 110°. Sheep have a field of vision of close to 360°, and when grazing they are moving their heads frequently and so can easily retain visual contact with other sheep. It certainly does not apply to sheep under all circumstances, for example for the sheep grazing the bush shown in Plate 2.4.

Leyhausen (1971) describes individual distance as the distance that gives a balance between dispersive and cohesive forces and "through individual distance, space itself becomes the vehicle of social meaning". He maintains that the distance between individuals depends on the rank of the respective animals, with greater distances occurring between animals of similar rank. However, there are no data published to support this quite feasible conten-

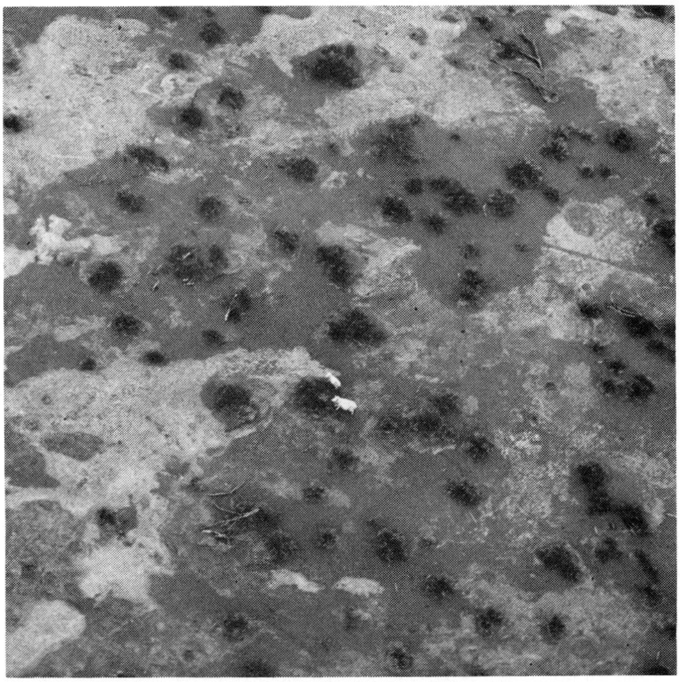

Plate 2.4. Merino sheep grazing bluebush in drought conditions in western N.S.W., Australia.

tion. Individual distance is that distance at which individuals will tolerate each other.

Social distance is that distance over which two individuals can maintain social contact, i.e. the distance over which signals can be passed. Baskin (1975) considers that the individual distance is usually two to three body lengths because he obtained values of 1.7 m in sheep, 3 m in reindeer, 5.1 m in horses and 7.7 m in camels. Baskin considers that there is another distance — which he calls the 'gregarious' distance over which animals of one group coordinate their movements with those of neighbouring groups. For this distance he gives 25 m in sheep and reindeer, 50 m in horses and 130 m in camels. Finally, he considers that there is a visual distance over which animals can identify others of the same kind and can achieve sexual contact. The figures (100—200 m for sheep, 500 m for reindeer and horses and 2500—3000 m for camels) are interesting but it is not stated how they were obtained.

Animal communication is an important area for study. Little is known about the types of signals used by animals in communication or the senses involved. Understanding these would help in understanding many aspects of social organisation. Most information is qualitative and concerns communication of rank or between mother and offspring. The latter will be discussed in Chapter 5. Most communication of rank and assertion of dominance is visual, and detailed descriptions of aggressive and submissive postures are available elsewhere for most domestic species. Since they are not specific to free-ranging conditions they will not be described.

Both sheep and goats can be aware of human presence at distances of a kilometer. This has been observed by McDougall (1975) in goats and by Arnold (unpublished) in sheep. It is not known whether sight or hearing has the greatest acuity at long distances. On balance it seems to be sight, provided the line of view is not obstructed. We have observed that sheep approached upwind will follow the movement of man from a distance of a kilometer. Geist (1971) found that Bighorn sheep followed movements of coyotes at a distance of 900 m. Response to sounds is tuned to unaccustomed noises.

The sense of smell has not been studied although McDougall (1975) found that goats could be approached to within 3 m upwind before sensing the presence of man. Sensitivity to odours downwind is much greater. Bighorn sheep are said to be able to scent a man at 300 m (Geist, 1971).

Social leadership and dominance

Other aspects of social organisation to be considered are spatial leadership and dominance. There have been few studies on either topic with grazing animals. In a study of groups of six sheep, Arnold (1977) found that the initiation of movements was by individuals that were both less gregarious and more independent, frequently grazing with their backs to other sheep or at a greater distance from other sheep than is usual. The movements of these sheep were 'followed' by other sheep and so they were 'leaders' but not in any positive way.

In other studies (Arnold and Maller, 1974) in which competition for supplementary feed was being studied, it was found that there was no correlation between the rank in which sheep reached the feed troughs and their competitive ability at the troughs. However, certain individuals were consistently amongst the first few sheep to reach the troughs. Thus, it appears that movements when grazing or going to feed are initiated by a few sheep and are followed by others — dominance is not involved. In feral goats Riney and Caughley (1959) found no signs of leadership even when billies were present. Hafez et al. (1969) state that in cattle leadership is passive, i.e. that individuals with lower cohesive tendencies gradually move away from their neighbours, and, if in the van, others will follow.

In horses, Grzimek (1944) found that leaders in movements were not the dominant animals. In the presence of a stallion, mares were observed to "graze and lie down in response to his direction, but when there is not a stallion present there is much less cohesion in the mares' activities" (Collery, 1974). This was not the case with a herd of eleven mares and a stallion (Arnold, unpublished) in which any activity could be initiated by any individual. Similar results were obtained in another study, although movement to water was generally initiated by one of three mares (Schoen et al., 1976). Movement between locations by Exmoor ponies is initiated by any animal, although if a subordinate pony is leading over a long distance it often stops and lets a dominant pony take over (Tyler, 1972).

Social dominance has rarely been observed to be exerted when sheep or cattle are grazing or are camped. It certainly is seen in horses where subordinate animals deliberately avoid moving too close to dominant animals and dominant animals frequently threaten subordinates (Arnold and Wallace, unpublished). A dominance hierarchy exists in flocks of sheep, and herds of cattle and horses. The hierarchy in cattle is relatively stable, changing little over the years between individual cows (Von Sambraus, 1977). It is exhibited by males during mating and in competition for supplementary feed given in a restricted space, or at water troughs in all three species. It appears also to influence the order of walking by cattle, as has been observed in dairy cows.

In the movement of cows between the paddock and the milking shed, those that are most consistently in the van are of medium dominance, with dominant cows in the middle of the line. Cows of low dominance tend to be in the rear of herd movements (Kilgour and Scott, 1959). The animals 'leading' the herd on any one day varied in this study, with 13 cows occupying the first five positions. The structure of the rear was more fixed with only 10 cows being found in the last five positions. A similar result was obtained by Beilharz and Mylrea (1963). Leyhausen (1971) describes this type of structure as one in which leadership is by pushing rather than by pulling.

Dominance when feeding at troughs in flocks of sheep of similar age is not related to weight (Arnold and Maller, 1974). In flocks of mixed ages the youngest and oldest are the least competitive and in flocks of mixed breeds Merinos are less competitive than crossbred sheep like Dorset Horn × Merinos. Under severe competition the poorer competitors cease trying to feed. The same general pattern has been observed in horses (Grzimek, 1944).

Wagnon et al. (1959) noted that in a herd of 2 5 year old Hereford cows fed supplements on range that the 2 year olds were driven away from the feed. After removing the 2 year olds it was noticed that the 3 year olds were pushed away. An analysis of the factors influencing dominance ranking in herds of 12 cows (Wagnon, 1965) showed that dominance rank was related to age in an asymptotic form, and linearly to weight and to aggressiveness, here measured as the number of contests an animal took part in. Agility of the cow was also stated to be an asset! In another study (Wagnon et al., 1966) a significant difference was found between breeds (Angus, Shorthorn and Hereford) in social dominance. The Angus was most dominant and the Hereford least. There was a *negative* relationship between social dominance and size between breeds, but within breeds there was a *positive* relationship.

Kilgour and Campin (1973) observed the behaviour of groups of bulls of different ages. Up to $4\frac{1}{2}$ years the bulls formed a herd with a dominance hierarchy and developed specific areas of a paddock for sparring. Older bulls did not form a herd but tended to prefer certain areas of a paddock and some individuals became strongly territorial.

Cows in a herd recognise their social rank relative to each individual. In dairy farm practice adjustments are made to the herd as cows calve, and in many situations herds are wintered in barns in groups and regrouped each spring to form a herd for grazing. Both of these disturbances cause social disruption.

A stranger introduced into herds of cows with a well formed social dominance hierarchy is frequently involved in fighting as it finds its place in the hierarchy (Schein and Fohrman, 1955). Bremner (1975) studied the effects of cows re-entering a herd after separation for <6 or 6—11 weeks. She found that fighting increased, as opposed to mere investigation by other cows, with longer periods of separation, viz. 5% were involved in fights when separated <6 weeks and 18% when separated >6 weeks.

Social interactions between species

There are interactions between species of animals when they are grazed in the same paddock. It is usual to see sheep and cattle grazing apart when in the same paddock (Plate 2.5) but cattle and horses may often be intermingled whilst grazing (Plate 2.6). The sheep may use the same area of a

Plate 2.5. Sheep, cattle and horses grazing in the same paddock.

Plate 2.6. Cattle and horses mixing together.

paddock as the horses and cattle but at a different time of the day; this is dealt with in detail later.

There is a clear dominance of horses over cattle and sheep (Tyler, 1972) and cattle over sheep should they come together, with horses displaying

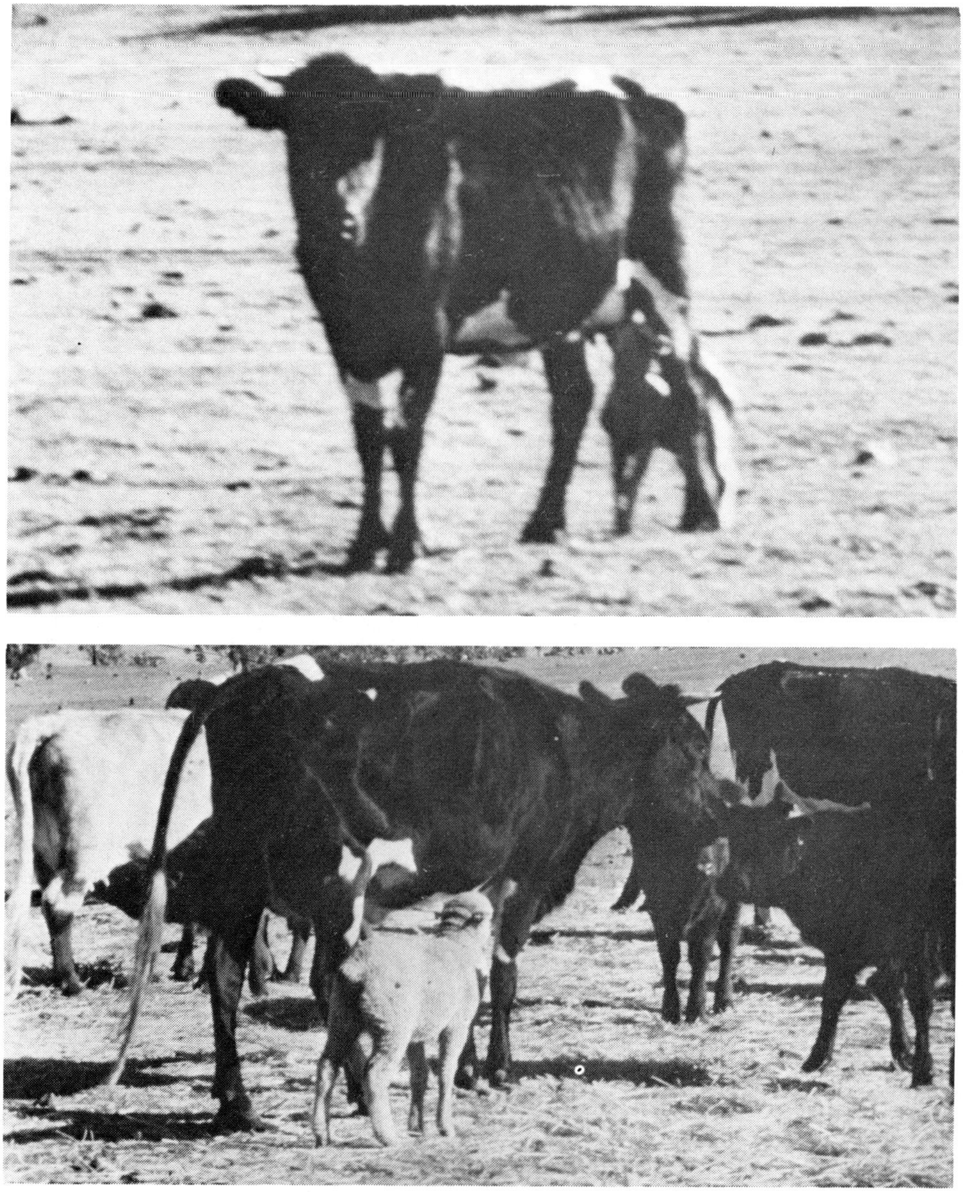

Plate 2.7. Friesian cow with Merino lamb which she raised.

aggressive threats to cattle that are in their way. Both calves and horses have been observed to kick young lambs that are in their way but adult sheep move aside. On the other hand horses will allow sheep to take supplementary feed from between their legs, but will drive cattle away.

Social cohesion between sheep and cattle may occur occasionally. Bond et al. (1967) found that a single steer and a single wether left in a paddock form a bond due to the wether staying with the steer. This bond occasionally remained when more animals of each species were put into the paddock. Plate 2.7 shows a lamb reared by a cow. This strange occurrence is more surprising when the difficulty the lamb would have in reaching the teat and sucking from it is considered. Unfortunately, there is no record of the age of the lamb when it was adopted by the cow. On another occasion a mare that was not in foal attempted to adopt a calf soon after it was born, driving away the cow whenever it approached. This behaviour persisted for 2 days. The calf made no attempt to suck from the mare.

EFFECTS OF THE ENVIRONMENT AND SOCIAL ORGANISATION ON ANIMAL DISPERSION

The discussion so far has been concerned with the social organisation as it affects relationships between individuals and thus affects formation of flocks or herds. The patterns described may, however, vary to allow the animals to adapt to the environment in which they live. This is rarely uniform in topography or vegetation; supplies of water may be close or far apart. Each of these factors influences dispersion of the animals in the environment and the behaviour of the flock or herd, as does the size of the paddock and the density of the vegetation. In addition, climatic factors can have a strong influence on dispersion.

Exploratory behaviour

When they are introduced to a new paddock, sheep and cattle explore it. Initially they move around the boundaries and follow this by moving further into the paddock. In a small flat paddock which allows clear vision of all parts, sheep will rapidly start to use all of the area. However, physical or visual constraints may mean that several days or weeks pass before they have explored and started to move through most areas of the paddock. It has been found by Arnold (unpublished) that young sheep introduced into a 300 ha paddock with tall grass and shrubs with which they had no previous similar experience, kept to within 20 m of the fence line for 5 days and failed to find water, which was further in, and so had to be driven to it. Sheep reared in a similar environment had no such problem. Thus experience is important and not only for sheep. Jardine and Anderson (1919) state that horses raised on plains do not adjust themselves to rugged mountain ranges.

Movement routes

Both the exploration phase and the subsequent daily pattern of use of an environment depend on the field of view of the animals. Cattle, being taller, have a much clearer field of view than sheep and this is reflected in their behaviour. In large paddocks sheep tracks usually run parallel to fence lines for up to 50 m, provided there are no obstructions, and then at some distance from the fence, run in various directions (Fig. 2.6, Plate 2.8). It appears that sheep use the fence lines for orientation. On the other hand, cattle tracks rarely run close to fences but in near straight lines between locations, e.g. from grazing or camping areas to water (Plate 2.9), under both intensive and extensive conditions. In undulating terrain they generally found the easiest routes of travel, as along the crests of ridges or in gaps between hills (Weaver and Tomanek, 1951), or parallel with contour lines. The cattle use a path until it is worn down or eroded and then start another parallel with it.

Cattle make fewer tracks than sheep because they usually travel in single

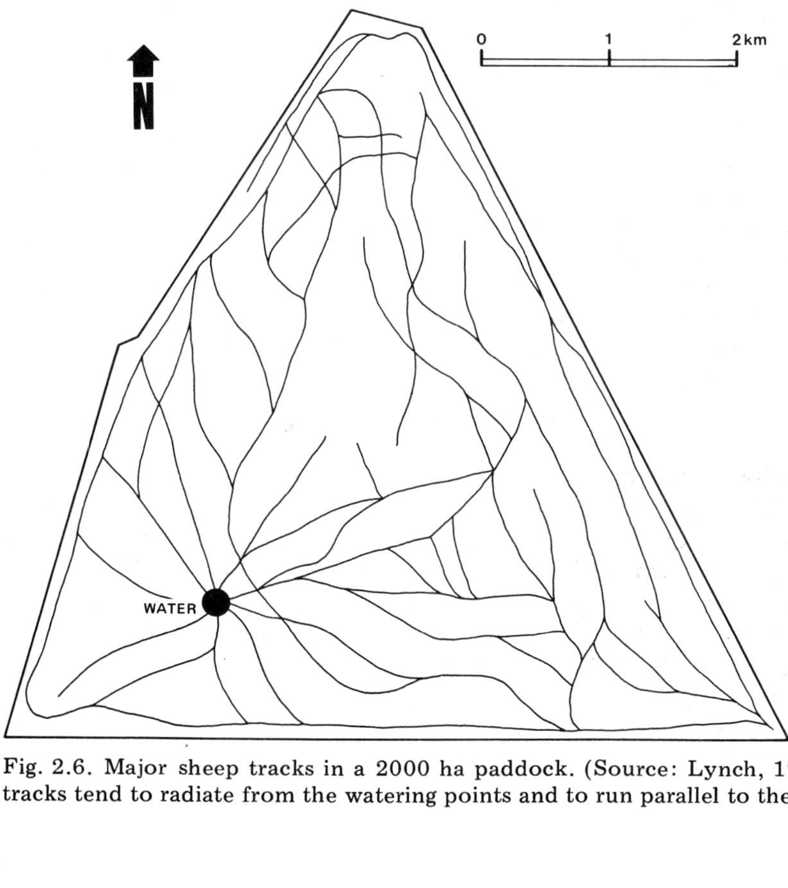

Fig. 2.6. Major sheep tracks in a 2000 ha paddock. (Source: Lynch, 1974b.) *Note*: The tracks tend to radiate from the watering points and to run parallel to the fences.

Plate 2.8. Sheep tracks on undulating country.

file (Plate 2.10b), whereas a flock of sheep on the move will often be in several parallel lines, as seen in Plate 2.10a. In large paddocks sheep may establish straight tracks radiating out into the paddock but they often start from watering points (Fig. 2.7).

These tracks of sheep and cattle starting from water end in gradual branching where feeding areas are located, and the animals fan out to graze. Sheep do not follow tracks whilst grazing on flat paddocks of pasture but many do so in hilly country, and on rangeland (Plate 2.11) where there may be a network of 'paths' around individual groups of small shrubs or tufted grasses.

Cattle do not have set routes when grazing on small paddocks but Weaver and Tomanek (1951) found grazing routes on rangeland up to 40 m wide (Fig. 2.8). They occur particularly in ravines, and may form a network over the range. Sheep will also form tracks for general movement about the environment and do more 'trailing' in large paddocks than in small paddocks even when feed is abundant.

Most sheep tracks made during movement between adjacent feeding areas on hilly terrain have a slope of 1—2°, but tracks to camping sites or through passes in hills may be much steeper. On one steep hill a track to the camping ground inclined 22° where the maximum slope was 25°. Cattle tracks in similar country were found to be 2—4° for grazing routes. No cattle tracks

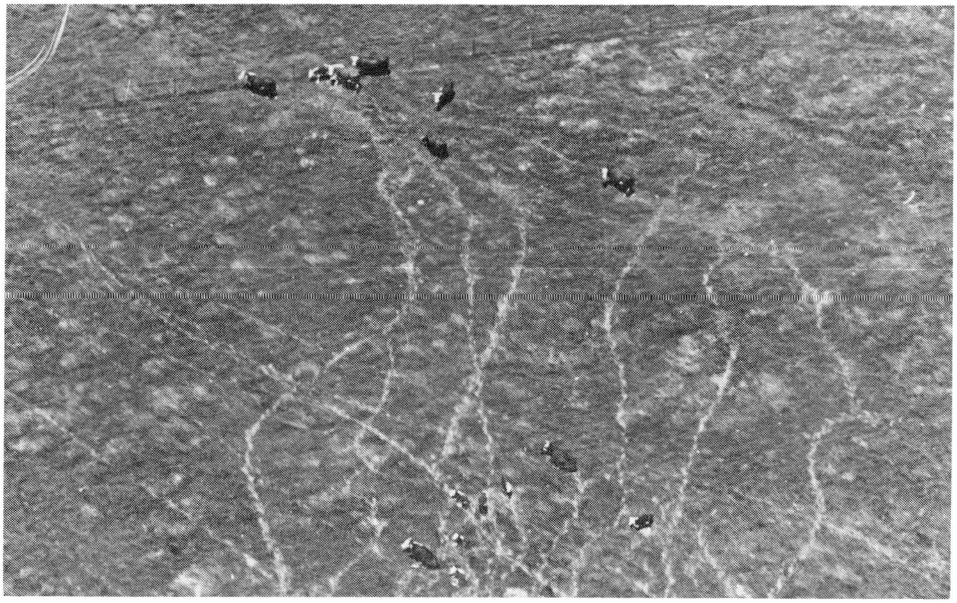

Plate 2.9. (a) Cattle tracks on pasture land in Southern Australia.

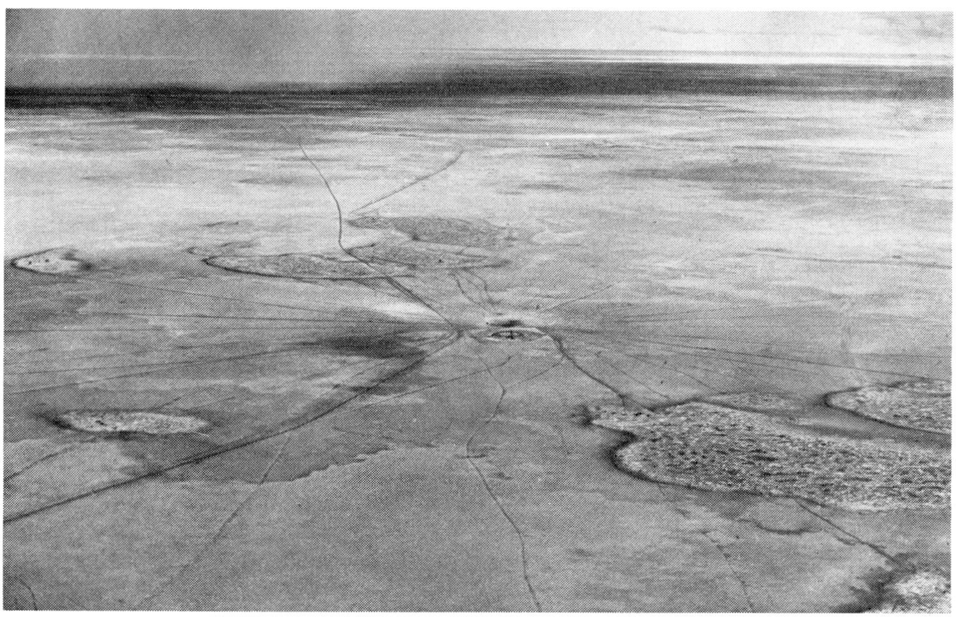

Plate 2.9. (b) Cattle tracks on rangeland in Northern Australia.

Plate 2.10. (a) Sheep walking.

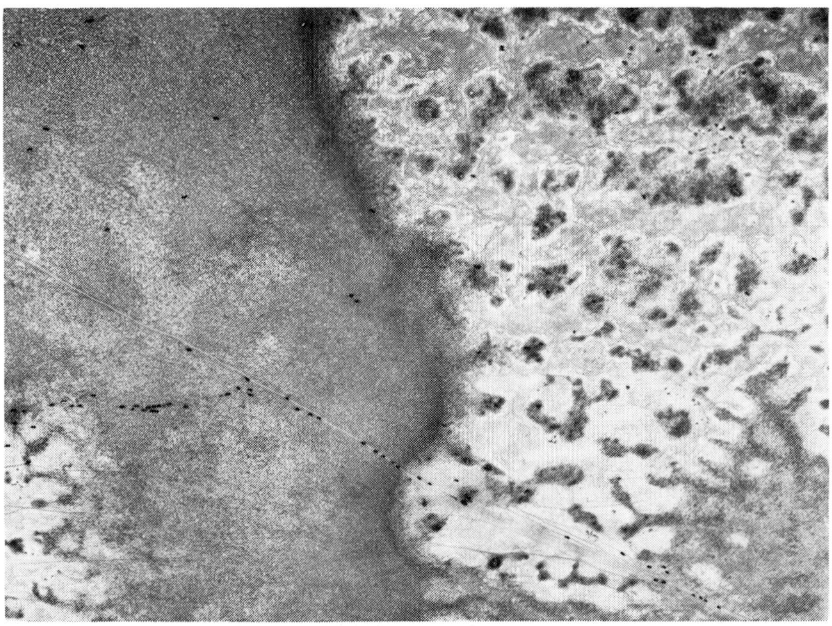

Plate 2.10. (b) Cattle walking.

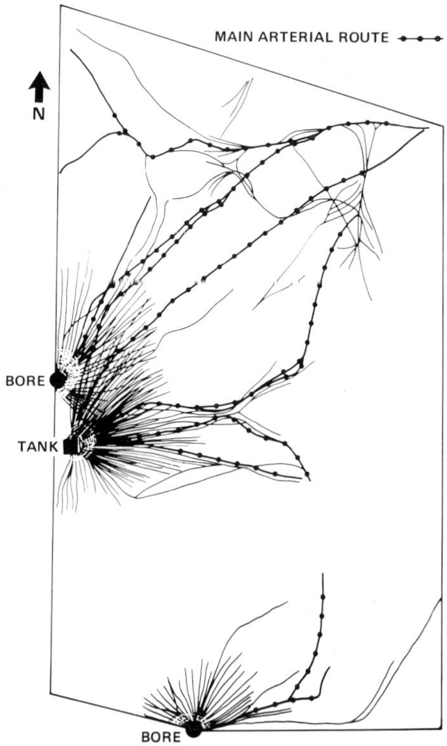

Fig. 2.7. The density and orientation of sheep tracks in a 1400 ha paddock of saltbush. (Source: Squires, 1974.)

were found going up hills, the steepest was 10° going for a short distance along a fence.

Factors influencing dispersion

Variations in the basic patterns of flocking in sheep have been discussed already — little is known about cattle or horses. However the sheep's basic pattern may be modified so that a breed like the Merino with a strong flocking or cohesive instinct may be split up into many sub-groups. The variables determining flock behaviour are age and breed, as discussed already, the topography and vegetation and the amount of feed available. Undulating country causes flocks to form groups; even light timber does. The degree of sub-grouping and the dispersion within these groups changes diurnally with more aggregation and decreased distances between sheep occurring just before and just after resting periods. Thus sheep will aggregate in large groups or as a whole flock when resting or when drinking and then gradually split up into smaller and smaller groups as they graze away from water or

Plate 2.11. Network of sheep tracks used when grazing on rangeland on the Riverine Plain of N.S.W., Australia.

camp sites (Squires, 1976). The grazing formation often begins as a wide arc or front (Plate 2.12) which then breaks up as the sheep move. The effects of various factors on dispersion in Merino sheep have been studied by Dudzinski and Arnold, 1967 and Dudzinski et al., 1969, 1978.

Of the breeds studied, the Merino sheep shows the least tendency to sub-group. Given flat terrain and abundant feed flocks of 500 adult sheep or more remain as a single group. Dudzinski and Arnold (1967) twice surveyed the size of groups and distance between individuals on two sheep stations in Western New South Wales during drought conditions. The stations had paddocks of 3000—8000 ha containing 500—2000 sheep. The average size of groups differed considerably between the stations (Table 2.6) but relatively less between sheep, either in different paddocks or between type of sheep, i.e. wethers or ewes with lambs, or with time of day. The condition of the range was poorer on the station with the smaller group size. However, the type of vegetation and its distribution were also different. A further survey was made on four paddocks at the Kayrunnera sheep station when forage conditions had improved considerably (Dudzinski et al., 1969). The average distance between sub-groups decreased from 33 to 22 m when range conditions improved and the average nearest neighbour distance decreased from

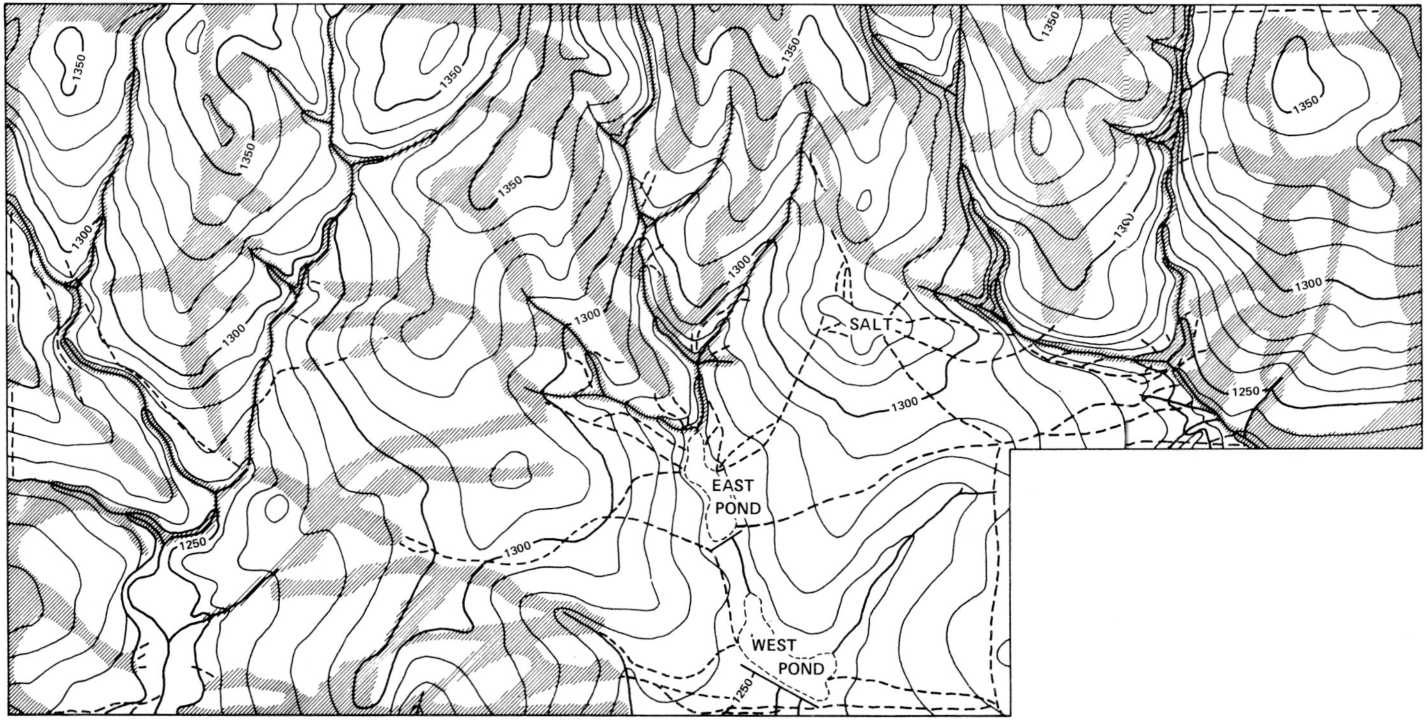

Fig. 2.8. Grazing routes of cattle in a hilly paddock. (Source: Weaver and Tomanek, 1951.)

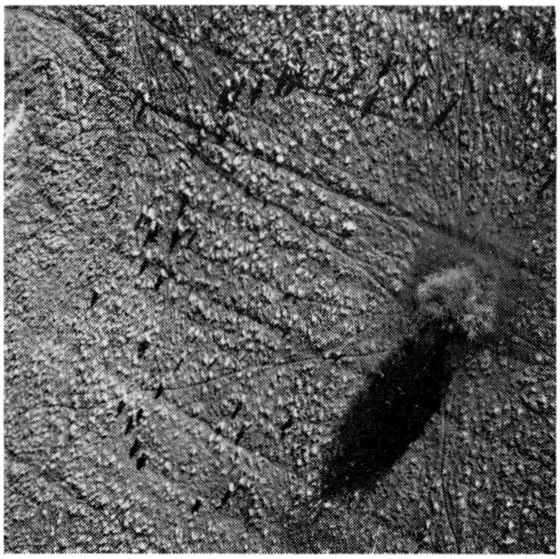

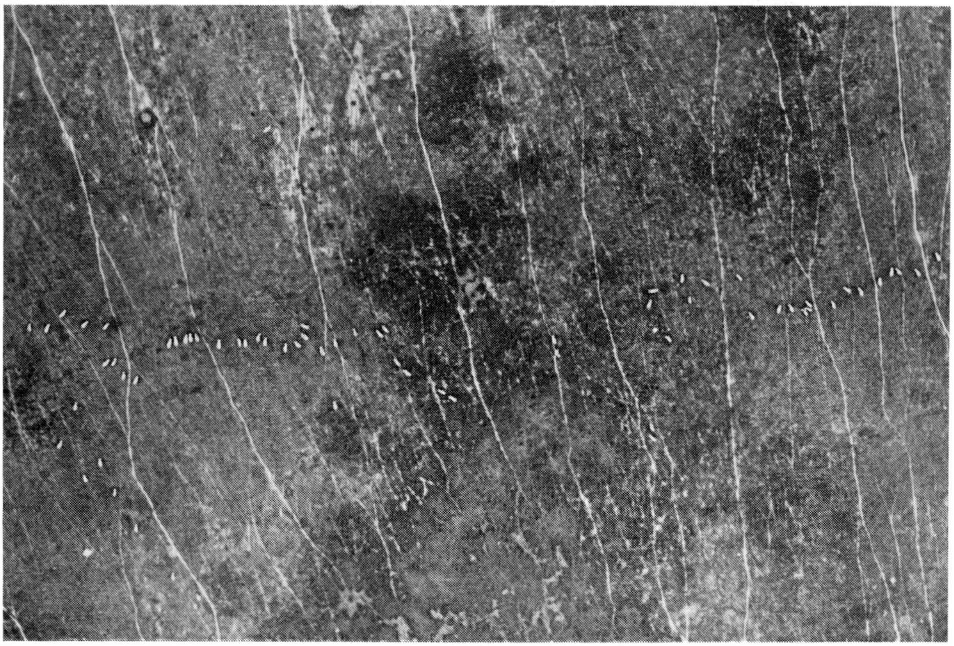

Plate 2.12. Sheep grazing formations. (a) (b) Grazing front formed when sheep have been camped. (c) Line formation.

5.5 to 4.3 m. This is illustrated in Fig. 2.9. Thus it seems that drought conditions forced the sheep to be further apart; the nearest neighbour distance was much greater than under abundant pasture conditions (2.3 m). It seems though, that groups will split up rather than spread out, thus maintaining as close contact as possible between individuals.

TABLE 2.6

Size of groups on two sheep stations in New South Wales and mean distance between individuals

		Kayrunnera		Mt. Murchison	
		Group size	Distance (m) between sheep	Group size	Distance (m) between sheep
Drought	August 1965	5.9	5.4	12.0	5.5
Drought worse	November 1965	5.7	6.7	13.3	5.5
Recovery from drought	May 1967	7.1	5.3	—	—

Type of vegetation had a small influence, the range in distances between groups and nearest neighbours varying by 10% between paddocks at any one survey, with some consistency in that one paddock always had the greatest distances for both measurements. The way the height and distribution of different species in the vegetation affects dispersion has been illustrated by Squires (1975b), and is shown for these two stations in Plate 2.13. The greater density of bushes at Mt. Murchison allowed more sheep to find suitable feed within close vicinity of one another than at Kayrunnera where groups of two sheep were found, often grazing on the same bush.

In another semi-arid environment, at Meekatharra, Western Australia, Gardiner et al. (1978) and Dudzinski et al. (1978) found that a flock of 330 Merino ewes grazing in a 5100 ha paddock gradually dispersed into smaller groups with deteriorating pasture conditions (Fig. 2.10). There is an exponential resistance to dispersion, indicating that size of group was also of critical importance in social organisation. Group size was related to body weight in this study, and when sheep were near death from starvation they were found either alone or in large groups that presumably had given up the task of foraging.

The mean values given for nearest neighbours do not tell the complete story because they hide the problem of one mean obtained from a number of different distributions found in the flock. Fig. 2.9 gives a comparison of distributions for the Merinos at Kayrunnera under drought and under near normal range conditions. Under drought conditions the Merinos' dispersion was similar to that of British hill sheep under normal conditions (Fig. 2.5).

Much less is known about 'sub-herding' in cattle (Plate 2.14). Cattle on unfertilised annual range in California split up into groups of two to three, but on range fertilised with sulphur remained as one herd (Green et al., 1958). Hereford cows were seen to graze in groups of 10—20 on rangeland in Oklahoma, but the composition of the groups was constantly changing

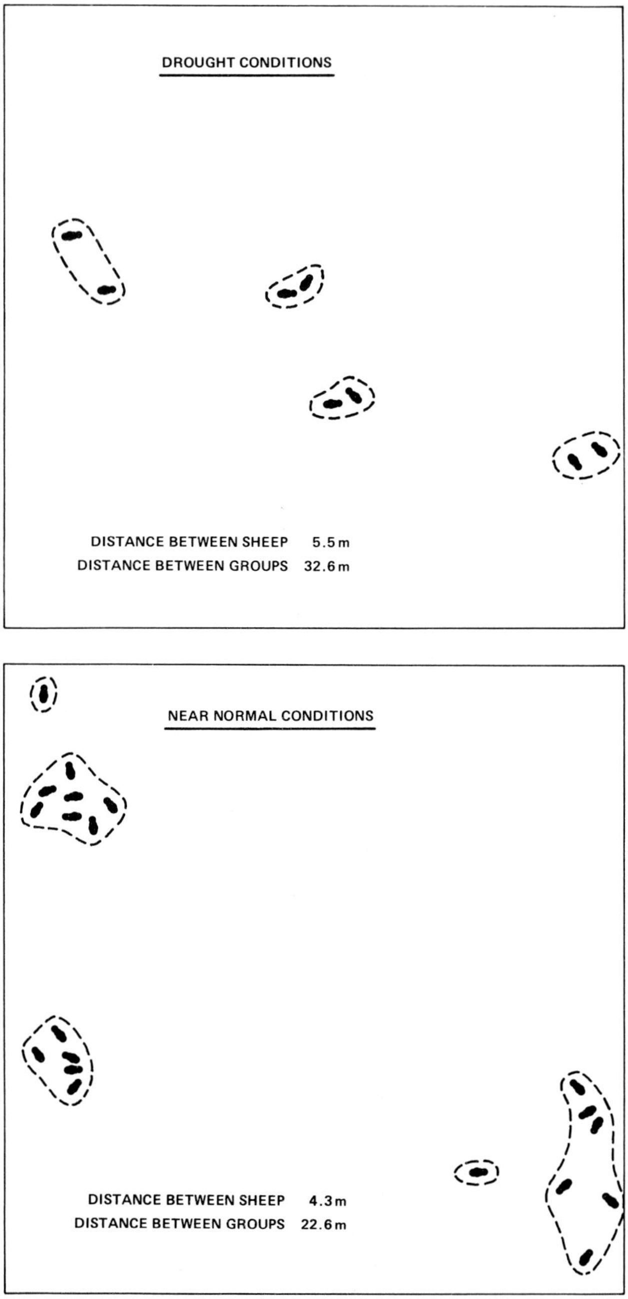

Fig. 2.9. Examples of patterns of dispersion under drought and near normal range conditions at the Kayrunnera sheep station, N.S.W., Australia. (Source: Dudzinski et al., 1969.) *Note*: The patterns are typical for one specific paddock and show that the average group size increased from 2 to 4 sheep with improved range conditions whilst the distance between groups and between individuals decreased. The dotted lines indicate groups of sheep.

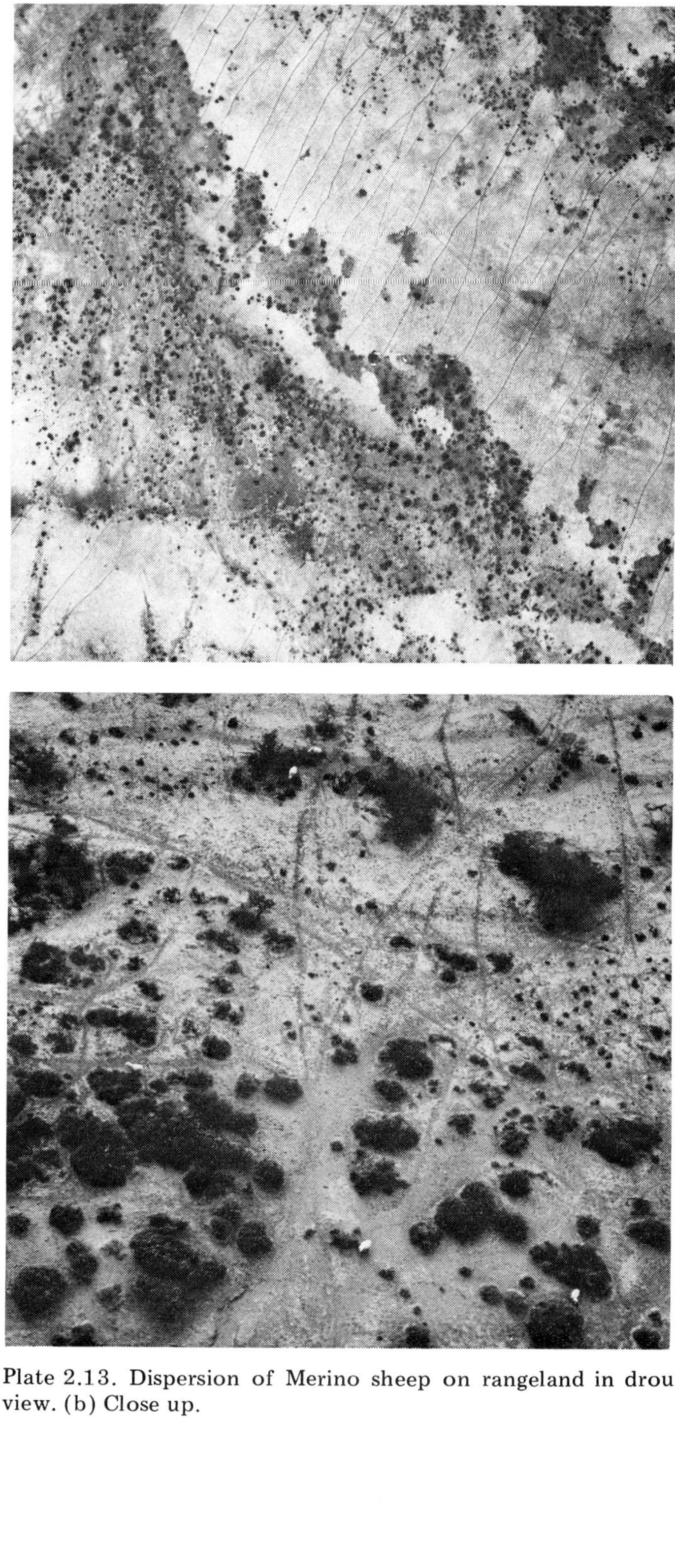

(a)

(b)

Plate 2.13. Dispersion of Merino sheep on rangeland in drought conditions. (a) Distant view. (b) Close up.

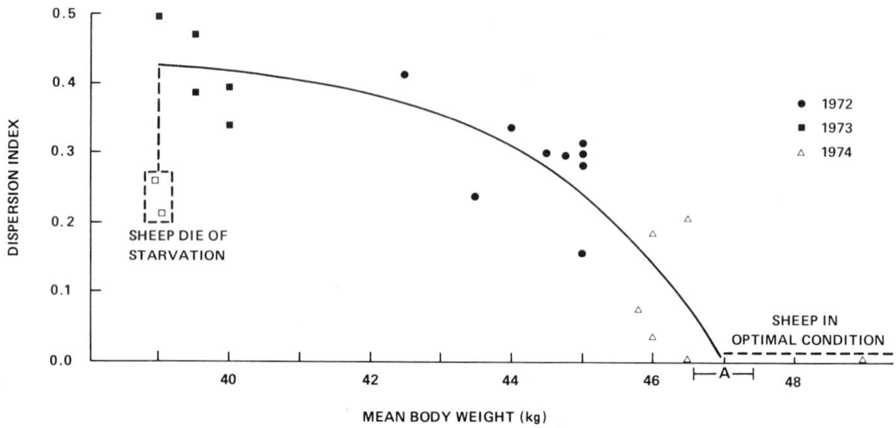

Fig. 2.10. Relationship between body weight and dispersion index in Merino ewes on range land at Meekatharra, W.A., Australia. (Source: Dudzinski et al., 1978.) *Note*: The dispersion into groups was measured by the Shannon-Wiener statistic and range condition was reflected in body weight. A is the threshold at which the range condition is too poor to allow the flock to stay as one unit.

Plate 2.14. Cattle herd broken up into sub-groups.

(Dwyer, 1961). Schmidt (1969) found that Shorthorn cattle in the Northern Territory split up into groups of <30.

The most detailed study has been with Shorthorn cattle grazing on a 153 km^2 paddock in Central Australia (Müller et al., 1976; Low, Dudzinski and Müller, unpublished). The herd of more than 300 cows broke up into small mobs even when feed was abundant. These varied in size from one to 85 animals, but over 5 years mobs of more than 80 were found only four times. The average mob size varied from 3.9 to 11.2 when the cows grazed six different plant communities. Within a plant community mob size varied more with the quantity than with the quality of the feed (Fig. 2.11).

Home range behaviour

In many pastoral situations sheep are moved from paddock to paddock as the farmer, rancher or pastoralist sees fit and the individual animals rarely adopt home ranges. The only recorded case of this occurring was by Lynch (1967) who found that a flock of 400 Merino sheep grazing a 325 ha paddock split into one group of 300 and six small flocks that ranged over areas that did not overlap. In the upland rangeland of Britain it is normal for sheep to have home ranges so that common land or unfenced mountain country can be grazed and the sheep will not wander from their own areas.

Firbank (1946) describes this well in his book about a Welsh hill farm. "Each ewe has her own beat, and will always be found near to the same grassy hollow or sheltered gully. She brings up her lamb on her little range,

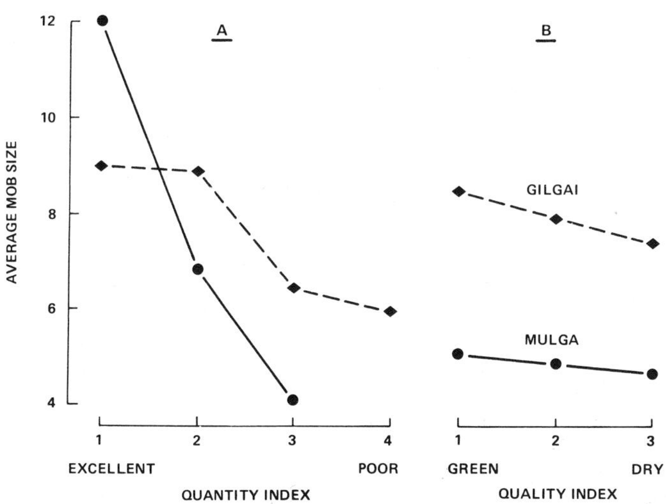

Fig. 2.11. Changes in mob size in Shorthorn cattle with changing range condition on two plant communities at Alice Springs, N.T., Australia. *Note*: Mulga is a savannah with scattered trees whilst the gilgai community is a flat area with no trees.

and when in time she is sold the lamb carries on the tradition, and eventually bequeaths the domain to her own offspring. This legacy, multiplied by a thousand individuals, ties the flock to its home. Sheep from adjacent farms naturally mingle along the common boundary, but if a dog is sent along the dividing line the sheep make for their respective sides like hairs parting before a comb."

Since Firbank wrote, home range behaviour has been studied in detail by Hunter (1964). He studied Scottish Blackface sheep on a heft (paddock) in the Cheviot Hills where they had been for living memory. He found that the sheep could be classed as belonging to one of three home ranges (Fig. 2.12) that were overlapping. He suggests that home ranges develop by social competition whereby some sheep are forced to graze in areas with less favoured plant communities. He demonstrated that when the stock numbers on a heft were reduced then the grazing pressure on the most preferred areas did not decrease because sheep previously grazing on less preferred areas moved in to replace sheep taken off the heft.

The restriction of sheep to a home range is strong. Rawes and Welch (1969) studied Swaledale sheep grazing on 3850 ha of moorland where 32 flocks are grazed in an unfenced area. They moved 30 sheep belonging to two distinct home ranges, one on the west side and one on the east side of

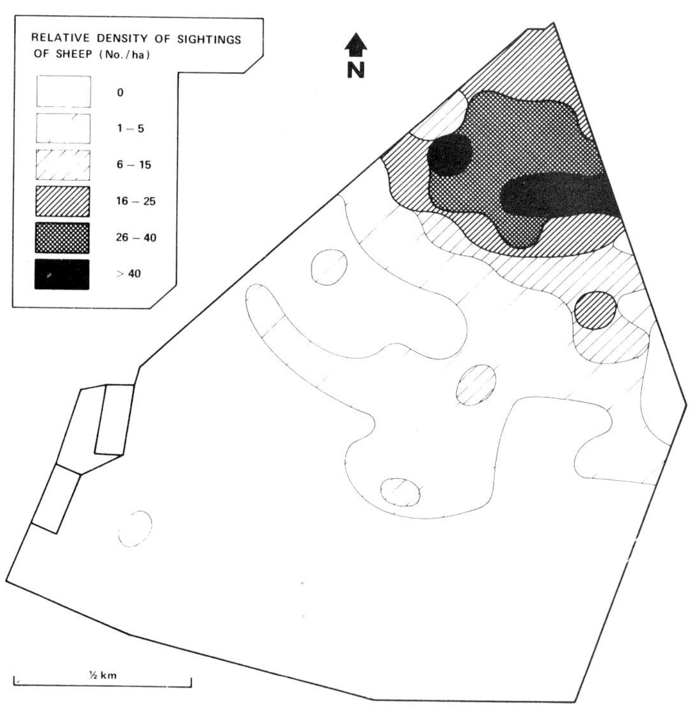

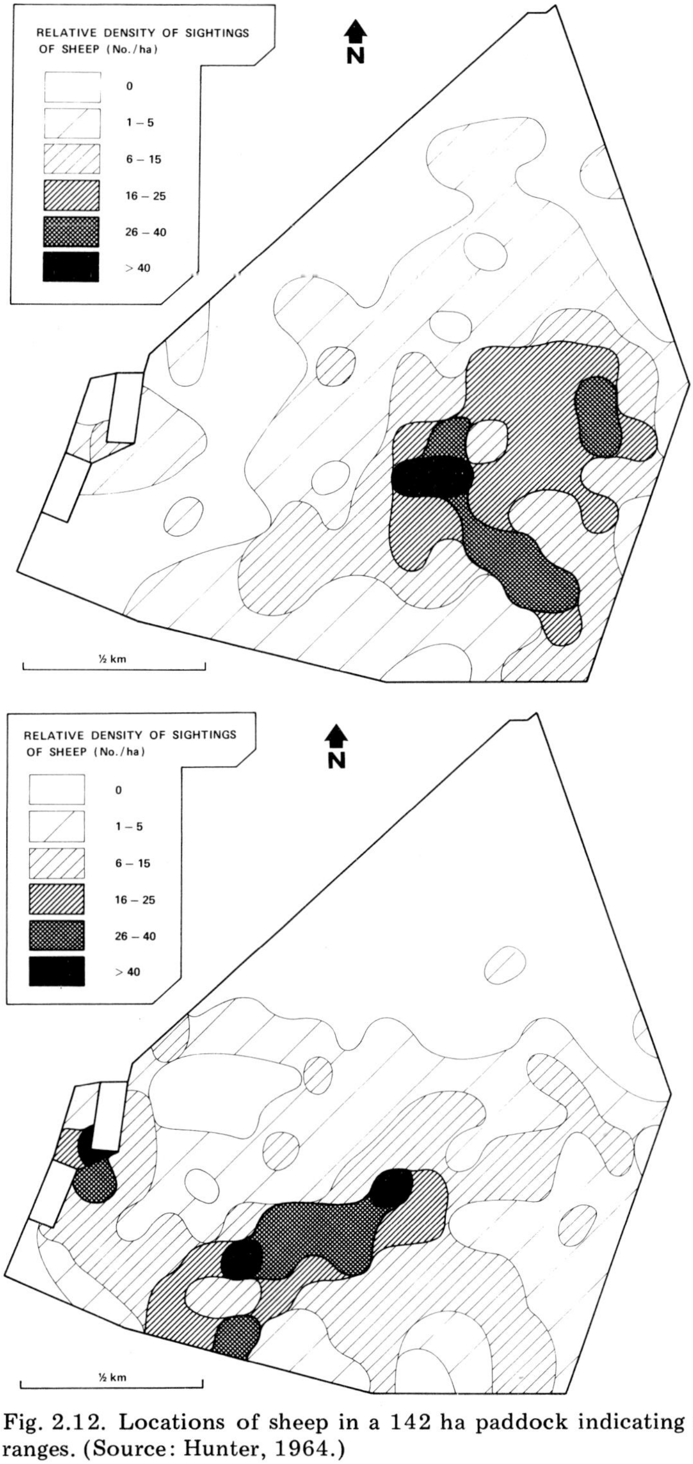

Fig. 2.12. Locations of sheep in a 142 ha paddock indicating grouping into three home ranges. (Source: Hunter, 1964.)

the area, to the west side. Three sheep from the east side walked 7 km back to their home range and stayed there.

Home range behaviour was shown also for South Country Cheviots (Hunter and Milner, 1963) and for North Country Cheviots (Griffiths, 1970). Griffiths found that 38 young sheep he studied in a flock of 174 split into two groups that rarely mingled. The whole flock also split into two groups. He substantiated two of Hunter's earlier findings. Firstly, that a new group of sheep, added to the flock when it was reduced to 73, occupied an area on the fringe of that used by the original flock; secondly, that sheep removed from the heft for 4 months went back to their original home range when returned to the heft. Hunter and Davies (1963) had found that removal of weaner sheep at 3 months of age for 3 months did not prevent the sheep remembering and returning to the home ranges where they were born. Separation for a longer period broke the adherence to their original home range.

Fayre (1976) reports that sheep from different farms occupy different home ranges when put on alpine pastures in the Hautes Alpes of France.

Horses in the semi-wild state are said to be territorial (Imanishi, 1950; Bruemmer, 1967), but it is more likely that they have home ranges as Tyler (1972) found with Exmoor ponies. These home ranges were 70—1000 ha in size and very stable. On the other hand Hall found that the size of the home ranges of the feral horses in the Pryor Mountains of Montana depended on the availability of forage, and water. They have separate winter and summer home ranges. Camels in the Sahara are closely attached to home ranges. Even years after separation from an area they have a tendency to return to their original home range and their owners may find them 1000 km away from where they were last seen.

Effect of weather

During the course of a day grazing animals are unlikely to use the whole of the available space, except under very intensive conditions. The area used will be influenced by weather conditions and also by factors that are, so far, undefinable. This is illustrated for sheep in Fig. 2.13. The two days were similar to the observer in weather conditions, yet the areas of the relatively small paddock used for grazing were quite different. It could well be that the pattern of use is almost random when weather conditions have no effect. It would be determined by movements initiated by various individuals during the day. This suggestion is made because, as is discussed earlier, there is no overt leadership in sheep flocks when they are grazing.

Movement in relation to wind has been mentioned before, i.e. that with light winds sheep and cattle, especially on hot days, move to the part of the paddock which provides most wind. However, the animals cannot continually graze into the wind or they would end up standing against a fence.

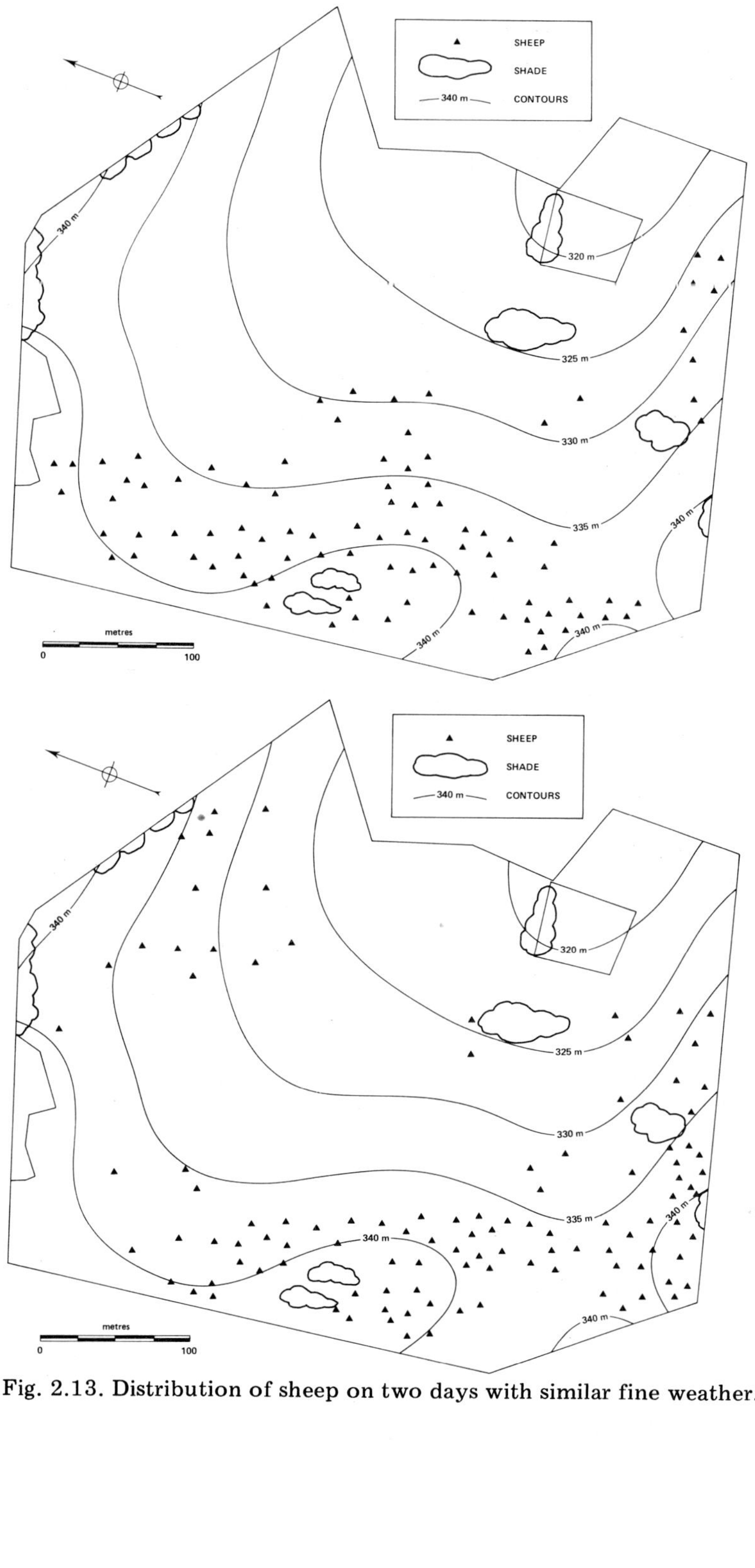

Fig. 2.13. Distribution of sheep on two days with similar fine weather.

TABLE 2.7

Orientation of cattle in relation to wind in summer in Oklahoma (Source: Dwyer, 1961)

	Per cent of animals		
	Side to wind	Head to wind	Head upwind
Cattle grazing	68.3	20.5	11.2
Cattle standing	73.1	16.8	10.1

Dwyer (1961) observed that cattle mostly graze and stand with their flanks to the wind (Table 2.7) and remain in the windiest part of the paddock. This has relevance to the diurnal pattern of movement by cattle. Woolfolk (1949) found that Rambouillet sheep grazing on rangeland in Wyoming were grazing into the wind 44% of the time, with the wind 30% of the time and sideways to the wind 26% of the time.

THE USE OF SPACE AND ITS EFFECT ON USE OF VEGETATION

In paddocks with a homogeneous type of vegetation grazed at a high grazing pressure, every square metre will be visited by the animals, probably every day. When feed is in short supply the herbage may well be grazed uniformly by sheep. However, this is an unusual occurrence with cattle, because they will avoid grazing on areas contaminated by faeces. At low grazing pressures even sheep will utilise only part of a paddock, concentrating their grazing close to water and camping sites. As heterogeneity of vegetation and topography increases, so does the variation in use of the environment by grazing animals. Variation in use is also influenced by the distribution of water sources and the distances animals will travel away from water to feed. Weather conditions also play a part. All these have important implications for pasture and rangeland management as is discussed in Chapter 6. Here we are concerned with the general patterns of behaviour.

Whilst there is no evidence to show that flocks or herds that split up will disperse further and so make better use of the vegetation, it may well be so.

Effects of location of water

On semi-arid and arid rangelands the location of water has a dominant effect on use of the vegetation. On these rangelands water is often restricted to one or two locations and degree of use of vegetation may decrease linearly or exponentially from a water source. Lange (1969) has termed the area around a water source a piosphere — from the greek 'pios' — to drink.

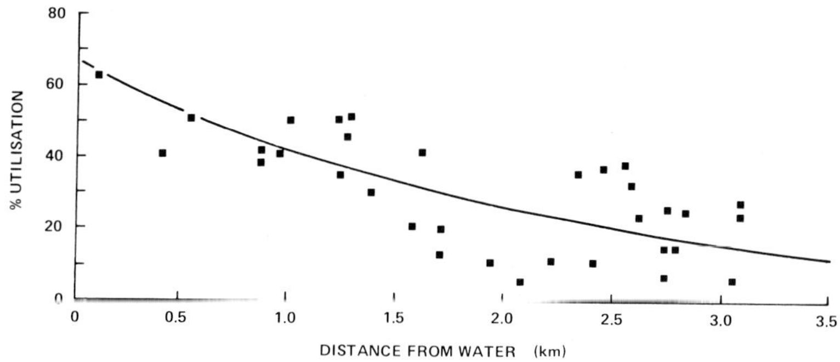

Fig. 2.14. Effect of distance from water on degree of use of vegetation on the New Mexico Agricultural Experiment Station, U.S.A. by cattle. (Source: Valentine, 1947.)

The shape of the curve relating vegetation use to distance from water will be influenced by type of vegetation, topography, season and class and age of livestock. Valentine (1947) studied cattle on a 1000 ha range in New Mexico and obtained the curve shown in Fig. 2.14. On this range, and on many cattle ranges in arid areas of Australia, there is a completely bare area of 0.5—2 km radius around the watering point. A smaller, but similarly denuded area occurs with sheep. The distance from water to zero utilisation of the vegetation by sheep in Australia is dependent on the type of vegetation. Thus Squires (1974) found that sheep regularly used only half of a 1400 ha paddock of saltbush and made no use of vegetation occurring at 3 km from water (Fig. 2.15), whereas distance to water did not limit distribution of sheep in a 2000 ha paddock which did not have any saltbush but where the furthest corner of the paddock was 4 km from the water supply (Lynch, 1974a).

In Central Australia, cattle herds may split up into two groups — the 'walkers' and 'non-walkers' described in Chapter 1. The 'non-walkers' forage close to watering points whereas the 'walkers' head straight from water to more distant feeding areas. However, there is a limit to how far they will go.

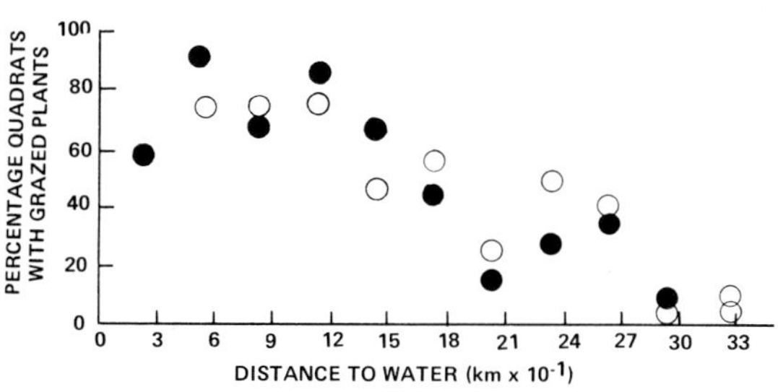

Fig. 2.15. Relationships between the occurrence of grazed *Chlaris* (solid circles) and *Sporobolus* (open circles) plants and distance to water. (Source: Squires, 1974.)

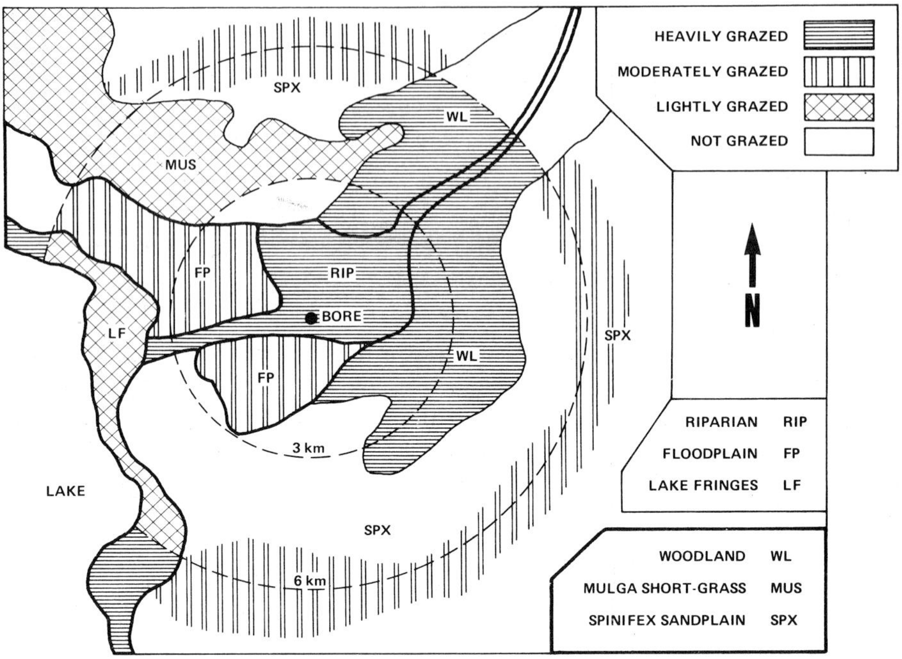

Fig. 2.16. Relative grazing use by cattle of plant communities around Sandy Bore, Napperby Station, N.T., Australia, during periods when the majority of cattle were grazing within 3 km, 3—6 km and >6 km from the bore. (Source: Hodder and Low, 1978.)

The majority of cattle graze within 4 km of water and only when forage conditions are poor do the majority of cattle graze beyond 8 km from water (Hodder and Low, 1978). Even so, they are selective in the types of vegetation they graze so that grazing intensity varies both with distance from water and with type of vegetation (Fig. 2.16). For example when the cattle are forced to graze more than 6 km from water they still graze preferentially even if this means a long distance is walked.

Effect of topography

The studies cited were mostly conducted on flat terrain. In hilly or mountainous country slope is an important factor influencing use. Mueggler (1965) studied 38 areas grazed by cattle in Montana. All were areas accessible only from the bottom of a slope. He found that with a 10% slope, 75% of the use was within 750 m of the bottom and with a 65% slope most of the use was within 30 m from the bottom.

In the Cache National Forest in Utah the effects of location of water and salt, the topography and type of vegetation on range use were studied (Cook,

1966). Twenty-one site factors that could influence use of a given site on the range were identified. Using a regression analysis Cook was able to account for only 56% of the variation in percentage use of grasses on a site. Over three years nine of the 21 variables contributed significantly to the regression. The most important were:

(1) % slope at the site
(2) % slope adjacent to water
(3) % slope from site to water
(4) distance to water below site
(5) % maximum slope between site and water
(6) % palatable plants on site
(7) thickness of brush around site
(8) % slope from site to salt.

Clearly, then, many factors influence the distribution of grazing animals and, despite the many studied by Cook, he was able to explain relatively little of the variability in use. Some other factors of importance are found from other studies.

In Central Australia, Low (1972) found that average cattle densities varied from 0 to 9 animals km^{-2} on the main plant communities, but seasonally the variation was much greater, with values up to 40 km^{-2} on one minor community. On the 153 sq km paddock the pattern of distribution was altered by localised rainfall, i.e. cattle moved to areas of fresh growth following localised rain.

In Nebraska, cattle on hilly rangeland were found to develop grazing routes (Weaver and Tomanek, 1951), as shown in Fig. 2.8. There was very considerable variation in use of the vegetation on different parts of the paddock, related to the location of these routes and to the direction of prevailing winds. A close correlation existed between location of the cattle and the direction from which the wind was blowing. Dwyer (1961) also recorded this and it is recognised in rangeland in Australia that sheep paddocks are most heavily utilised in the areas from which the prevailing winds come.

Effects of heterogeneity of vegetation

Vegetation condition, even where the range is of one plant community, will vary over the seasons and this influences animal distribution. For example Dwyer (1961) recognised three classes of land in the range he was studying (Table 2.8). The large contribution of the shallow land in July was attributed to a flush of growth of annual plants there at that time. Fertilizer practice can be deliberately used to change distribution as Smith and Lang (1958) showed in mountain country in Wyoming. They put out strips of nitrogen fertilizer on sites usually poorly utilised and cattle distribution was changed both in the year of application and in the following year. An increase in use of slopes on annual range in California was obtained by using

TABLE 2.8

Changes in distribution of cattle on rangeland in North Osage County, Oklahoma (Source: Dwyer, 1961)

Site	Percentage of total area	Percentage of grazing		
		June	July	August
Loamy prairie	70	77.1	51.4	74.2
Rocky breaks	18	14.6	9.9	7.4
Shallow	12	8.3	38.7	18.4

sulphur fertilizer (Green et al., 1958). Stewart (1953) obtained a similar effect with sheep in Scotland by using trace elements.

All three studies reflect changes in the acceptability of herbage grown after fertilizer was applied. They indicate that the same plant species have different acceptabilities when grown under different nutrient regimes. Such differences will occur naturally on different types of soil, as demonstrated by Cook (1959) in Utah. It was also a contributing factor in the degree of use made of a plant community in a study in North Wales (Hughes et al., 1975).

These workers examined sheep numbers found on 56 sites representing six broad types of grassland in the mountains of Snowdonia for more than 15 years. They found (Tables 2.9 and 2.10) that the overall use varied from 0.5 to 18.1 sheep ha^{-1}. However, high rainfall sites (>2450 mm) averaged 2.91 sheep ha^{-1} compared to 5.91 for lower rainfall sites — the sheep simply preferred not to graze where it rained so often! The same communities growing on basaltic and acidic soils had a threefold difference in sheep use, and not all of this was due to differences in productivity. They also found that sites with impeded drainage had heavier use than freely drained sites. These last

TABLE 2.9

The effect of rainfall and vegetation type on sheep population density in North Wales (Source: Hughes et al., 1975)

Vegetation type		Sheep population density at:	
		High rainfall (sheep ha^{-1})	Medium rainfall (sheep ha^{-1})
I	Herb-rich *Agrostis—Festuca* grassland	4.99	10.65
II	*Agrostis—Festuca—Nardus* grassland	4.40	6.40
III	*Molinia—Juncus* communities	2.12	5.44
V	*Vaccinium—Calluna* heath	0.34	2.89

TABLE 2.10

The effect of altitude on sheep population density in North Wales (Source: Hughes et al., 1975)

Vegetation type		Sheep population density at:	
		High altitude (sheep ha^{-1})	Medium altitude (sheep ha^{-1})
II	*Agrostis—Festuca—Nardus* grassland	3.11	4.40
VI	*Agrostis—Festuca—Cryptogramma crispa* communities	0.37	0.99

two factors are presumed to be due to differences in acceptability of the same species on the differing sites.

Detailed studies of rangeland use by sheep in the Cheviot Hills and the Pennines of England are relevant to this topic. Hunter (1962) studied South Country Cheviot sheep on a 102 ha hillside paddock in the Cheviot Hills on which there were nine types of vegetation. He found that 25% of the area sustained 51% of the grazing, with the different types of vegetation having a threefold difference in average use. Although the relative acceptability of the types of vegetation changed little over the year, there was a threefold difference between sites in use of a particular type of vegetation. Reasons for this include soil type and the particular admixture of the species common to the type of vegetation. Later work in the same area (Griffiths, 1970) showed a tenfold difference in use of the types of vegetation.

In Westmorland, a detailed study has been made of the use of a 3850 ha area of rangeland grazed by Swaledale sheep (Rawes and Welch, 1964, 1966, 1969; Welch and Rawes, 1966). Seasonal changes in the way sheep used one type of vegetation on three topographically different locations — a steep east-facing slope, the fell top, and a summit ridge — were studied first. The ridge had uniform use throughout the period (May—September) whilst use of the other two locations varied threefold. Wind had a marked effect on usage, with use being much reduced when winds were strong (>24 km per hour), so that sheltered sites were more heavily used. The east-facing slope had the heaviest overall use because it was in the lee of prevailing westerly winds.

A further study of the most favoured type of vegetation (Table 2.11) in three locations showed how sheep concentrated on the small areas of this *Festuca/Agrostis* grassland within large areas of other types of vegetation and that these were grazed at up to 18 times the intensity of the surrounding areas. However, the intensity of use of the *Festuca/Agrostis* depended on its location. In the upland area its use was reduced to nearly the same intensity as that of the surrounding area. Since the overall stocking rate of the range was 1.9 ha^{-1}, it is clear that the *Festuca/Agrostis* areas supported a large

TABLE 2.11

Density of Swaledale sheep on *Festuca/Agrostis* grassland and on surrounding areas on the Moor House National Park, Cumbria (Source: Rawes and Welch, 1966)

Site	Area (ha)	Sheep per hectare	
		Surrounding area	*Festuca—Agrostis*
Steep west facing escarpment	98	2.7	7.4
Upland above 600 m	51	2.0	2.7
Blanket bog	98	0.7	13.3

proportion of the sheep. In further studies Welch and Rawes (1966) found that grazing intensity on areas of blanket bog varied inversely with the heather content in the bog, from one sheep to 2 ha to one to 25 ha.

On Dartmoor, sheep and cattle and ponies were shown to use the different types of vegetation quite differently both overall and seasonally (Perkins et al., 1970). In August, for example (Table 2.12) cattle and ponies made no use at all of the bracken-invaded grassland and little of the *Vaccinium* moorland, whereas these two types of vegetation were the most favoured by sheep. Conversely, cattle and ponies made greater use of the short grassland and valley bog than sheep.

Effects of choice of grazing and camping sites on the return of nutrients

Even on cultivated pastures sheep, cattle and horses living together make different uses of the area. This is illustrated in Figs. 2.17 and 1.22. The greatest difference in grazing use of the paddock was between horses and cattle on the one hand and sheep on the other. Resting at night in summer and winter occurred mainly on higher ground and often in the same area of the paddock for all species, but nearly always adjacent sites would be chosen

TABLE 2.12

Relative use of types of vegetation on Dartmoor by sheep and cattle in August (Source: Perkins et al., 1970)

Type of vegetation	Sheep	Cattle + ponies
Upland bog	29	16
Short grassland	144	330
Bracken-invaded grassland	364	0
Vaccinium moorland	305	27
Valley bog	73	138
Calluna—Ulex heath	186	21

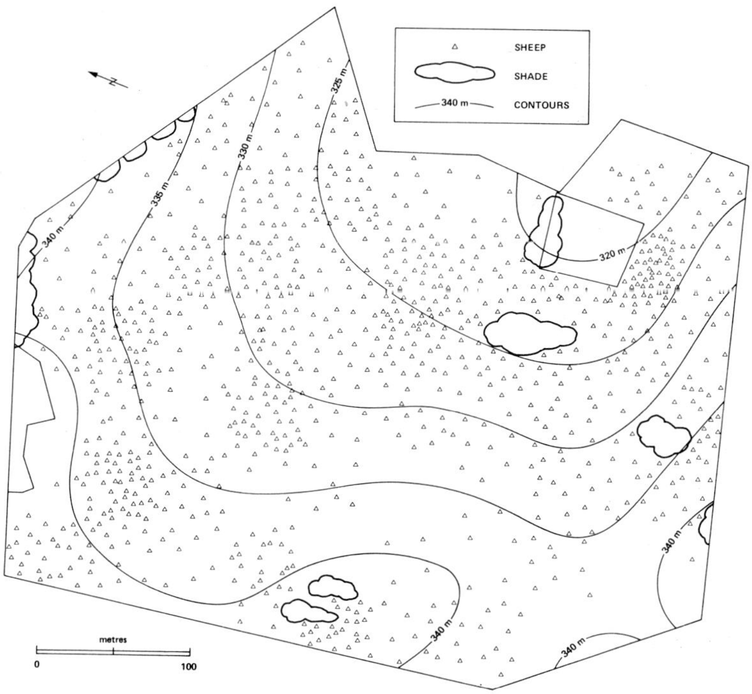
SHEEP
SHADE
340 m
CONTOURS
325 m
330 m
335 m
340 m
320 m
340 m
340 m
340 m
metres
0
100

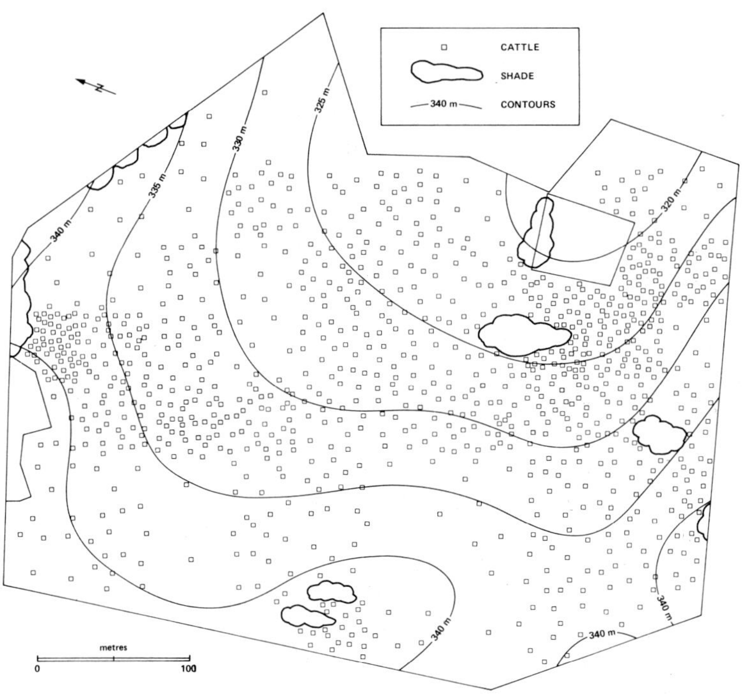
CATTLE
SHADE
340 m
CONTOURS
325 m
330 m
335 m
340 m
320 m
340 m
340 m
340 m
metres
0
100

Fig. 2.17.

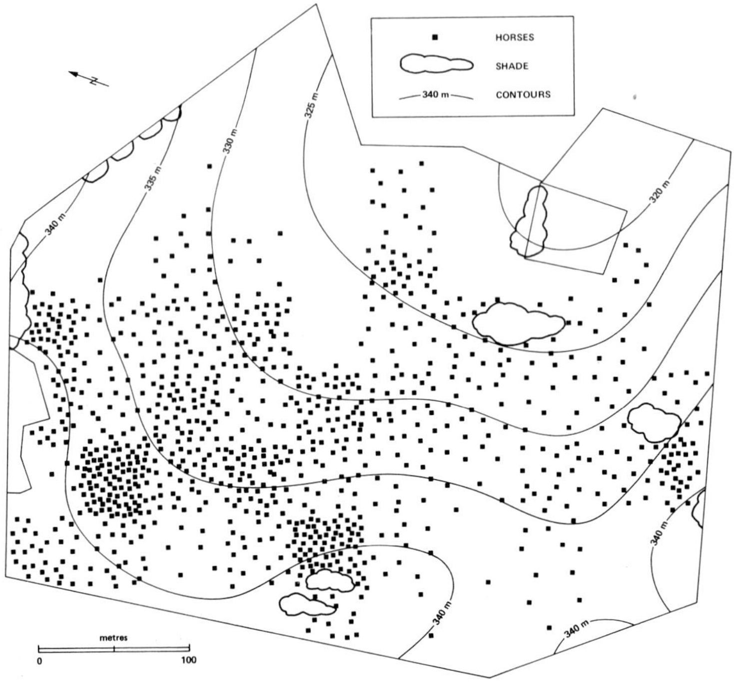

Fig. 2.17. Grazing use of a paddock by sheep, cattle and horses at Baker's Hill, W.A., Australia.

and camping would not be in mixed groups. In summer cattle and horses used trees for shade, but did not share shade, whilst sheep camped mainly in the open.

The return of nutrients via faeces and urine by grazing animals is rarely random. That voided when the animals are grazing may be fairly uniformly distributed when the animals are intensively grazed, as is often the case with dairy cattle. Under less intensive conditions the return will be far from random. The faeces and urine voided when the animals are grazing will be to areas preferred, and so, for example, on the moorlands of Britain the favoured *Festuca—Agrostis* grassland will receive a high proportion of the return by sheep.

Even in small paddocks (0.4—3 ha) faeces are concentrated around areas of special attraction, e.g. around water troughs, gates, fences, shade and shelterbelts (Petersen et al., 1956). Weather has its effect since it influences where animals graze and rest. Powell (1967) observed cattle in a 5 ha paddock and found that they moved to sheltered areas with increasing wind speed and defaecated there. The nutrients voided when the animals are at rest result in a considerable concentration of the return, especially where the animals have few resting sites.

In certain situations in Australia sheep have mixed camping sites, either on the top of hills in winter, or under trees that give shade or on bare areas around dams or water troughs in summer, and faeces are concentrated at camping sites. On the Northern Tablelands of New South Wales in paddocks ranging in size from 0.4 to 35 ha, 10% of the paddock may receive 40% of the faeces (Hilder, 1966a). On small paddocks in England a distinct non-random distribution of faeces has been found (Crofton, 1958; Donald and Leslie, 1969). In other situations the concentration is not so marked. Fig. 2.18 shows the distribution of return of faeces to a paddock in the south west of Western Australia grazed by sheep, cattle and horses during summer.

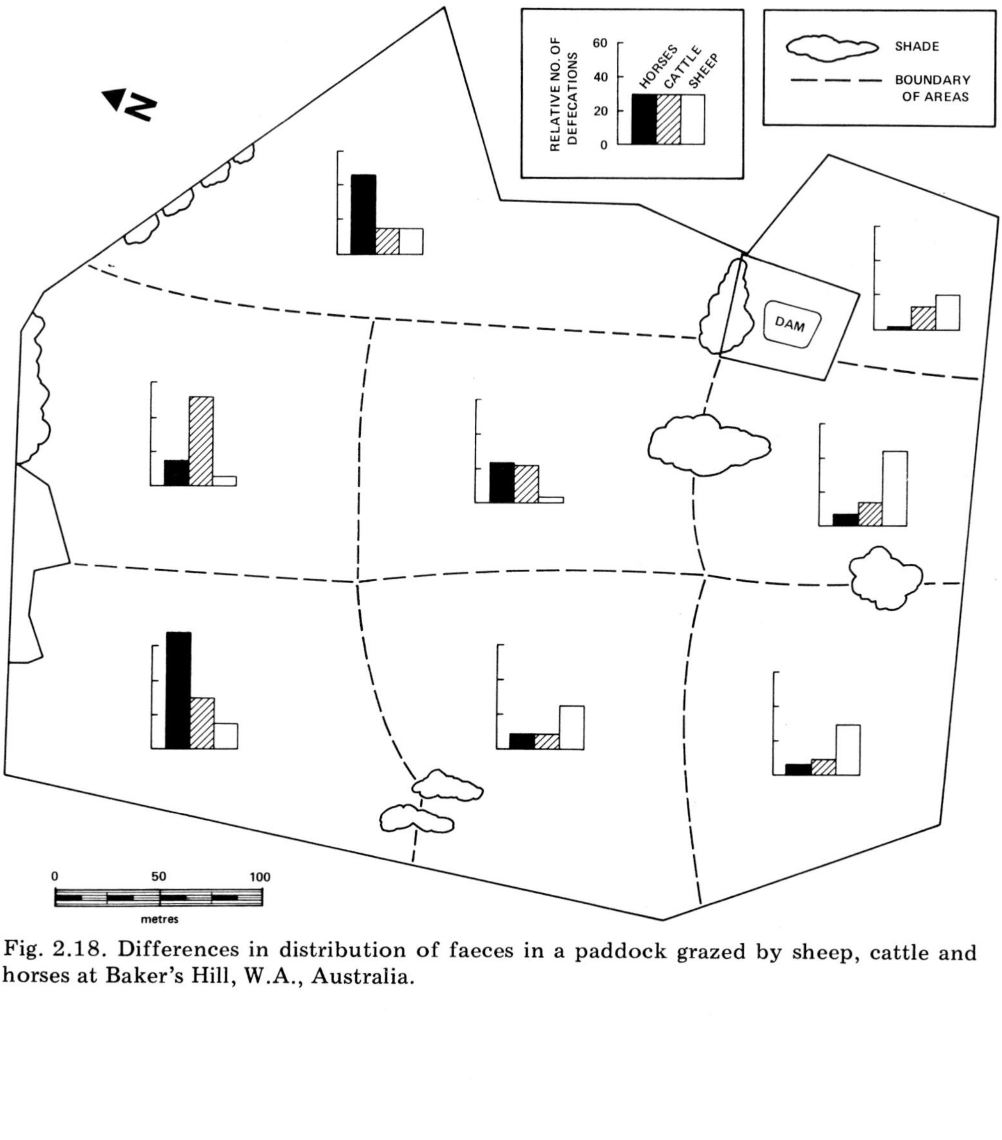

Fig. 2.18. Differences in distribution of faeces in a paddock grazed by sheep, cattle and horses at Baker's Hill, W.A., Australia.

Cattle produced the most uniform return and sheep the least uniform. Horses in this situation were intermediate.

In some parts of the world horses develop voiding areas in a paddock which they habitually use. This was noted by Hammond (1944), Taylor (1954) and Archer (1972) and has been examined in some detail by Ödberg and Francis-Smith (1976, 1977) in small paddocks. In these small paddocks horses do not graze the areas where they defaecate which soon become rank with long grass. If the stocking rate is high the paddocks develop areas of tightly grazed pasture and ungrazed areas. The rank areas are left ungrazed not because the herbage is unpalatable but because the horses, like cattle, dislike grazing close to where they have deposited faeces.

Why this pattern of behaviour occurs in some situations and not in others is not known. For example it is not seen on Exmoor (Tyler, 1972), nor on an Irish mountain farm (Collery, 1974). It may be a function of paddock size and/or of the stocking rate. With more space and a lower stock density the animals have the opportunity, and do, in fact use only part of the available space on any one day. Faeces dropped in the area may have lost their offensive odour by the time the animals return to the area so that rank ungrazed areas do not develop.

The pattern may also be influenced by whether the paddock is grazed by stallions or mares or by both. Stallions tend to mark the eliminations of mares with their own urine and faeces, as Bulbenik (1961) observed with groups of Przewalski horses in the Prague Zoo. This would tend to concentrate faeces. Many stallions regularly defaecate on the same spot, which again would concentrate faeces. Mares, however, do not normally do so.

The distribution of urine by cattle is similar to that of faeces according to Petersen et al. (1956), and this is to be expected. However, since both sheep and cattle preferentially graze herbage growing where urinations have been, no wastage of vegetation occurs as a consequence.

Chapter 3

DIET SELECTION AND FOOD INTAKE

In the previous chapter, it was stressed that different plant communities on rangeland are normally subjected to different grazing intensity, as a result of preferential selection by the animals. Within the communities two types of preference are seen: some species are preferred to others, and certain plant parts are found preferable to other parts. This pattern applies also in pastures with few species. Before considering the extent to which grazing animals are selective when grazing, and how selection is determined it is necessary to consider the mechanics of grazing.

THE MECHANICS OF GRAZING

There are several important differences between sheep and cattle in the way they graze. Sheep either bite the vegetation off the plant or it is broken as they grip it with their teeth and jerk their heads forwards or backwards. Measurements made from cine film of grazing show that 79% of the head jerks are backwards. Cattle use their tongues to gather vegetation into the mouth before biting and tearing it off, unless the vegetation is very short. Horses bite close to the ground (Plate 3.1), having the advantage over sheep and cattle of upper and lower sets of incisors. All three species move with their muzzles in a horizontal plane as they graze, and select in a vertical plane.

Because sheep have smaller mouths and teeth they can take smaller bites and so they are able to be more selective than cattle or horses if they wish and can bite closer to the ground. However, both sheep and cattle are able to vary their methods of harvesting food according to the structure of the vegetation. On lucerne, sheep will run the stems through their mouths stripping off the leaves (Arnold, 1960b), leaves being selected in preference to stems (Fig. 3.1). Cattle nibble leaves from stems in tall stands of grass (Stobbs, 1974). Stobbs showed that the amount taken per bite varied; from 0.05 g organic matter on tall mature tropical forage of low density, where animals nibbled individual leaves, to 0.71 g on swards of young ryegrass where the animals were almost non-selective.

The horizontal movement of the grazing animal's head results in a mower effect, with the crown of a plant being 'trimmed' off. This is a simple mechanical action and is not selective grazing.

Plate 3.1. Sheep, cattle and horses grazing.

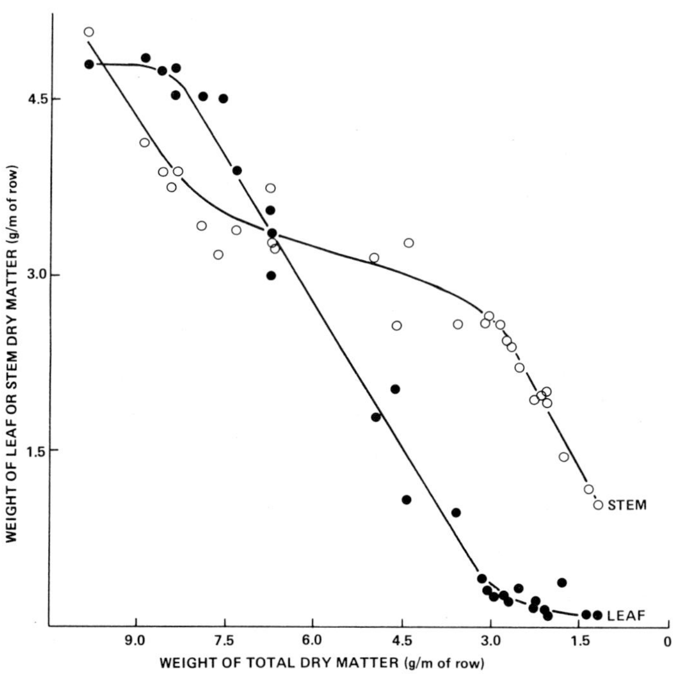

Fig. 3.1. Changes in the relationship between stem and leaf weight in a stand of *Medicago sativa* as a result of selective grazing by sheep. (Source: Arnold, 1960b.)

The frequency with which a plant or area of a pasture is grazed is related to the pattern of grazing. Under intensive rotational grazing the frequency will be several times a day, but under a set stocking or continuous grazing management this frequency is related to stocking rate and forage characteristics. The data from several studies with sheep were analysed. The frequency with which an area was grazed was found to decrease with increase in length and bulk of herbage, but to increase if there is a bulk of short herbage rather than a little long herbage, because the animals can eat less per bite. It also increases, of course, with stocking rate. Also the physical capacity of the sheep is limited; an individual sheep probably bites over 15—30 m^2 a day depending on the amount it obtains per bite.

However, not all areas of a pasture are visited with equal frequency, especially when food is abundant. Both sheep and cattle often restrict their grazing to certain favoured areas, thus creating heterogeneity in the vegetation even in pastures of a single species.

SELECTIVE GRAZING

The grazing animal grazes selectively and a whole range of factors interact to determine the composition of its diet on a particular day. These factors are summarised in Fig. 3.2. The composition of the diet it chooses may be quite unrelated to the proportions of various species or plant parts present and available to the animal. For example on a semi-arid rangeland in N.S.W., Australia, Leigh and Mulham (1966a, b) found that 80% of the diet during one period of the year was from plants that contributed only 1% of the total forage available. Similar situations have been frequently reported on rangelands where many species are available and feed is abundant in relation to stocking rate. On the other hand, on pastures with only a few species, of similar acceptability, the diet may closely reflect the relative availability of the species (Fig. 3.3).

There is general agreement that from the single plant, sheep and cattle eat leaf in preference to stem (Cook and Harris, 1950; Reppert, 1960; Arnold, 1960b, 1964a), and green (or young) material in preference to dry (or old) (Stapleton, 1934; Milton, 1953; Cook et al., 1956; Cowlishaw and Alder, 1960; Reppert, 1960; Arnold, 1964a). There may be exceptions to these basic rules, but they are rare. The material eaten, when compared with the material offered and rejected, is usually higher in nitrogen (Hardison et al., 1954; Cook et al., 1956; Weir and Torell, 1959; Arnold, 1960b; Bredon et al., 1967; Galt et al., 1969; Bedell, 1971; Coleman et al., 1971; Wallace et al., 1972), phosphate (Staten, 1949; Plice, 1952; Cook et al., 1956), and gross energy (Cook et al., 1956), but lower in 'fibre' (Hardison et al., 1954; Weir and Torell, 1959; Bredon et al., 1967; Coleman et al., 1971; Wallace et al., 1972). Opinion varies on whether eaten material is higher in sugars and minerals.

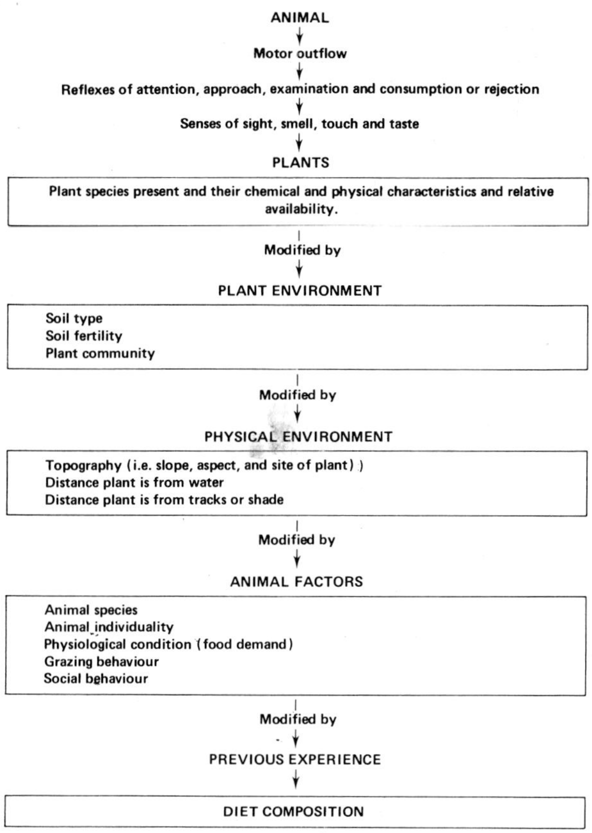

Fig. 3.2. Diagramatic representation of factors influencing diet composition. (Source: Arnold, 1964b.)

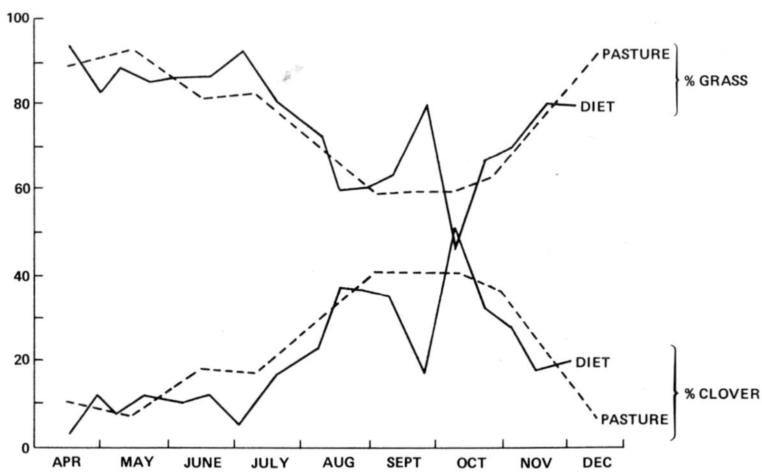

Fig. 3.3. Seasonal changes in the botanical composition and diet of sheep grazing a grass and clover pasture. (Source: Arnold et al., 1966.)

Other questions remain largely unanswered. Are green material and leaf material eaten because they contain higher nitrogen, etc., or simply because these parts are easier to harvest? Is there definite motivated selection for this material and, if so, is this linked to a nutritional wisdom? These questions will be considered in detail later.

THE ROLE OF THE SPECIAL SENSES IN SELECTIVE GRAZING

The senses of sight, tough in the lips, taste, and smell are all involved in diet selection.

Sight is used primarily to orient the grazing animal to other sheep, and to its environment. Sheep do recognise conspicuous food plants by sight but do not use sight to help them graze selectively. These facts were established by Arnold (1966a) who found that blindfolded sheep ate similar diets to sheep that could see under a wide range of conditions. Table 3.1 demonstrates this situation: sheep had a choice of eight types of pasture on which to graze. The scores for degree of defoliation were very similar for a sward type, whether sheep could see or not.

The senses of touch, taste and smell are important in diet selection (Arnold, 1966b; Kreuger et al., 1974). Arnold, in a series of studies with sheep treated surgically to produce single or multiple sensory impairment, showed that not only were there marked changes in acceptability of plant species, or strains of a species, when each of these senses was impaired but that total food intake may be affected, as will be discussed later.

In Table 3.2 the effects of impairing one sense on the choice made by

TABLE 3.1

Effect of sight on selection of green herbage by sheep in pastures containing mainly dry herbage (Source: Arnold, 1966a)

Pasture species	Scores of defoliation [1]	
	+ Sight	— Sight
Phalaris tuberosa	6.5	5.5
Medicago sativa	9.5	8.5
Trifolium pratense	5	6
Festuca arundinacea	9	8.5
Eragrostis curvula	0	1
Dactylis glomerata	6.5	7.5
Stipa hyalina	5.5	5.5
Lolium perenne	4.5	4.5

[1] 0 = no defoliation; 10 = complete defoliation.

TABLE 3.2

Percentages of strains of each plant genus that had similar and different acceptabilities to sheep with sensory impairment compared with acceptabilities of the strains to sheep with all senses intact (Source: Arnold, 1966b)

Genus	Sense impairment								
	Touch			Taste			Smell		
	I		0	I		0	+	—	0
Festuca	11	14	75	8	23	69	11	19	70
Dactylis	24	19	57	22	17	61	21	26	54
Phalaris	5	12	83	12	12	76	6	27	67
Bromus	14	42	44	15	25	60	9	26	65
Poa	25	16	59	33	10	57	18	16	66
Agropyron	19	8	73	19	7	74	19	7	74
Trifolium	22	25	53	54	0	46	40	4	56
Mean	17	20	63	23	14	63	18	17	65

+ Increase; — decrease; 0, no change.

sheep amongst 20 strains of each of seven pasture species is shown as *increase*, *decrease* or *no difference* from the ranking of the species by sheep with all their senses intact. The effects of removing any one of the three senses were generally the same, i.e., the ranking of a third of the strains was altered. Inability to taste improved the acceptability of more species than did the inability to smell or to feel them. A similar result was obtained with sheep grazing on a mountain pasture containing 15 species (Kreuger et al., 1974).

One of the most notable effects of depriving sheep of the sense of smell is that they no longer ate the flower heads of the grasses. Thus it can be concluded that the flower heads have an attractive smell.

Preferences were studied at various stages of growth of the pastures but there were relatively few strains in any species that consistently increased or decreased in acceptability when sheep were deprived of a particular sense. This is because the sensory stimuli received from the plant change as its chemical composition changes with stage of growth. An example is *Medicago sativa* (Table 3.3). In early spring sheep without the sense of smell ate much more of this species than did sheep that could smell, but by late spring the difference had disappeared and no difference occurred in summer.

The chemical signals which mainly influence food selection are those received at receptor sites for taste and smell. Stimuli are transmitted to the brain and the animal responds behaviourally or physiologically to the messages they contain. The animal responds by integrating these messages with others, such as feedbacks telling of the current nutritional state of the animal

TABLE 3.3

Differences in acceptability of *Medicago sativa* between sheep with the senses of smell and taste impaired and sheep with all senses intact (Source: Arnold, 1966b)

Date of test		Smell impaired	Taste impaired
Early spring	10. 9.63	+3.0 [1]	+1.0
	16. 9.63	+3.5	+3.0
	18. 9.64	+3.0	+1.0
	19.10.63	+1.5	−0.5
	24.10.63	+1.0	0
	30.10.63	+1.0	0
Late spring	30.10.62	0	0

[1] At each date the acceptability was assessed as scored on a 0—5 basis.

or the presence of some metabolic disturbance. The desire to eat may result in lowering either taste (Goatcher and Church, 1970) or smell thresholds of rejection.

The only signals that can activate this system are molecules which react chemically on receptors that transmit information. Therefore, in looking at food selection in relation to chemical composition of plants, only molecular concentrations of specific chemical forms are of significance. This hypothesis of communication implies that animals are able to distinguish significant stimuli from a background of chemical 'noise'. Thus, responses are given by the receptors only when receiving stimuli which are significant in that particular environment. What will be 'noise' to one species could be key information to another, through either a differing sensitivity or a completely different response. The relative intensity of a signal will be determined not only by the number of molecules received, but also by the number of receptors that can receive them. Differences probably occur between and within animal species not only in the total number of receptors but also in the ratio of, say, bitter to sweet taste receptors. Thus, different animals of the same species can be expected to differ in their taste response in given situations. The effect of this is to produce differences in selection between animals in a herd or flock. These differences are discussed later.

Just how the sense of smell operates in selective grazing is not clear. We hypothesize that ruminants can discriminate odours only at short range and that they are thus able to make precise discrimination between adjacent plants and plant parts. This hypothesis is based on our own experimental evidence including an important observation noted with cattle. Diet samples were being collected from the cardia of rumen fistulated cattle grazing pasture which had many 'dung patches'. That the sense of smell was used by these cattle was established because of the marked audible variation in inhalation intensity as the animal grazed. It was observed that deep inhalations were made when the muzzle passed over 'dung patches' and these were left uneaten, but immediately a 'dung patch' had been passed inhalations were no longer audible and the cattle grazed. With sheep, in studying the social bonding of ewes with their lambs, we have also found that odour discrimination operates only over very short distances.

What is known of the olfactory and gustatory responses of ruminants to various pure chemicals has been described by Arnold and Hill (1972). There is no doubt that failure to understand how ruminants select their food has led to enormous wasted effort in seeking 'palatability factors' in pasture plant species.

There have been numerous attempts to relate preferences to the proximate composition of plants. But it is not possible for the animal to recognise such things as nitrogen, 'crude fibre', 'energy', silica or 'ash'; these fractions do not exist in this form at molecular level in the plant. Where correlations are found between any of these characteristics and preferences, they must relate to specific compounds or some physical property of the plant. In the case of ash, it could be related to the content of sodium or potassium salts; with fibre, the ease of harvesting could be the significant factor (Evans, 1964). Animals cannot select a food because it is high in 'energy' or produces the best 'weight gain'.

When analysing reports in the literature it is important to recognise that a lack of correlation between preference rating and the content of a single chemical compound is not proof of the unimportance of that compound in determining preferences. Because of the multidimensional nature of the process, all that is indicated is that it is not of paramount importance. Conversely, a significant correlation between a single compound and preference is not proof that it is the main component influencing preference, unless all other components have been accounted for.

In considering the relationship between the chemicals influencing taste and the chemical composition of the plant, little attention seems to have been given to the chemical form of the components in the plant tissue, or to the proportion of these making contact with the senses. Most results are based on the total concentration of a component, or group of components, frequently determined on oven-dried material where the chemical form of the components will be quite different from that in the living tissue. The influences of plant and saliva enzymes are not considered and are largely unknown, a major problem being to account for the differences in the amount of chewing by individual animals at the time of ingestion. Both type and amount of enzymes and amount of chewing could well determine the chemical forms present at the taste-buds, and thus could be the deciding factor in acceptance or rejection.

SOME BROAD RELATIONSHIPS BETWEEN CHEMICAL COMPONENTS IN HERBAGE AND HERBAGE ACCEPTABILITY

Sugars

It is highly unlikely that an animal could give an integrated response to 'soluble carbohydrates'. What this fraction includes depends not only on the particular carbohydrates present in the plant under consideration, but also on the methods by which they are extracted and determined. Most techniques, while adequate for energy considerations, are unsatisfactory for sweetness measurements. It may seem more likely that the amounts of hexoses and pentoses could be related to preferences on the basis of sweetness, but then these may be present as a wide range of stereoisomers of mono-, di-, tri- or polysaccharides, glycosides or other derivatives, each of which gives a different sweetness response. Even pure glucose in solution occurs in seven interconverting forms, each with its own specific stereochemistry.

The type of sugar can influence ruminant preferences. Calves are known to have a much greater preference for sucrose solutions than for solutions of other sugars at a given molarity (Kare and Ficken, 1963) and it seems that sheep prefer glucose to sucrose at high concentrations (Arnold and Hill, 1972).

Despite reservations about considering a gross indicator like 'soluble carbohydrates', it may still be a useful index of preferences in the absence of better guidelines. The same applies to other broad classifications of compounds.

Significant positive correlations between preference rating and 'soluble carbohydrate' content (ethanol soluble) have been obtained with sheep feeding on varieties of cocksfoot (Reid et al., 1966, 1967), but not with sheep feeding on varieties of tall fescue (Reid and Jung, 1965).

Bland and Dent (1964) compared the preferences of cattle for 14 strains of *Dactylis glomerata* at two locations at several times of the year. For early spring growth at one site, preferences were most closely correlated with per cent 'total sugars', with hexoses and with fructoses but not with sucrose content. At all other observations 'total sugar' content and preferences were not related.

A similar result was obtained with grass sorghum (Gangstad, 1964) in that preference rating of cattle was correlated with per cent 'total sugar' but not with per cent 'soluble carbohydrate' (in this case mainly starch). He concluded that cattle like high levels of 'total sugar' and low levels of starch.

Cattle would eat unacceptable 'dung patch' herbage when it was sprayed with sucrose or molasses (Marten and Donker, 1964). Thus, for cattle, objectionable stimuli can be suppressed or overridden by the presence of sugar.

Since there are no reports of decreased preferences with increased sugar content, it seems likely that sugar content is important in determining ruminant preferences; but then sweetness is not confined to sugars, and certain sugar forms can give responses other than sweetness for non-ruminants.

Organic acids

The only published work (Jones and Barnes, 1967) showed a close positive correlation between sheep and cattle preferences and the citric plus shikimic acid contents of eight temperate grasses. This was repeatable over two periods of a year.

Tannins

Tannins are known to depress digestion in the rumen and to be closely related to preference rating amongst strains of *Sericea lespedeza*. Wilkins et al. (1953) demonstrated a 70% drop in voluntary intake of *Sericea* herbage by cattle when tannin content increased from 4.8 to 12.0%. Steer preferences amongst 1206 strains of the species were related to tannin content and to stem fineness (Donelly, 1954).

Coumarins

To humans, coumarin has a sweet smell and bitter taste. The smell of coumarin is objectionable to sheep (Arnold and Hill, 1972). Ashton and Jones (1959) sprayed ryegrass—white clover herbage to give it 0.5—1.0% coumarin. They observed that sheep, whilst initially (2—3 h) avoiding the sprayed herbage, eventually consumed it. The difficulty with such a study is that a single extra component is added to the complex taste and odour of the herbage. Since ryegrass—white clover herbage is usually very acceptable to sheep, it may be interpreted that the level of coumarin was insufficient to alter markedly the acceptability of this herbage.

One of the coumarins, *o*-hydroxycinnamic acid, which occurs in *Melilotis* spp. is said to lower the acceptability of the herbage of this species, but there is no evidence of this.

Isoflavones

In the genus *Trifolium*, many species contain isoflavones in quite large quantities. In *T. subterraneum*, an important pasture plant in Australia, up to 10% of the plant dry weight may be isoflavones in the free form and as isoflavone glycosides (Hill and Arnold, unpublished). Preferences of sheep amongst nine lines of *T. subterraneum* over two growing seasons were related to neither total isoflavones nor to amounts of any individual isoflavones. Also, Francis (1973) found that a 'no-isoflavone' mutant of a high isoflavone strain had the same acceptability to sheep as the parent strain. Thus, although these isoflavones can cause serious reproductive disorders in sheep, sheep show no discrimination against strains of clover containing them.

Alkaloids

A number of alkaloids in various alkaloid classes are known to reduce acceptability of strains of several food plants. In most cases, these alkaloids also have physiological effects on ruminants, either through impairing digestion in the rumen, or directly on the mammalian system. Probably the discrimination in selection of alkaloids represent the clearest case of preferences being physiologically wise.

In the genus *Lupinus*, some 30 different but related quinolizidine alkaloids occur, both sweet (alkaloid-free) and bitter (alkaloid-containing) strains occurring within species. The main alkaloid in *L. angustifolius* is lupanine, which can account for as much as 2.5% of the herbage dry matter (Gordon and Henderson, 1951). Sheep avoid grazing lines of this species containing alkaloids, if at all possible, but readily graze alkaloid-free varieties. According to Gladstones (1970), sweet and bitter strains of a species differ only by a single gene so that other chemical components may be similar. If this is so, then alkaloids may well be the only factor influencing acceptability.

Pyrolizidine alkaloids are found in several important herbage species particularly *Lolium* spp. and *Festuca arundinacea.* The main alkaloid is perloline, but festucine may also be present. As yet, there is no evidence that these affect acceptability, although perloline depresses digestion in the rumen (Bush et al., 1972).

Considerable attention has been given to relationships between the indolealkylamine alkaloids and acceptability of strains of *Phalaris arundinacea.* It was initially found that low acceptability was associated with a water soluble chemical compound (Roe and Mottershead, 1962; Kennedy and Holgate, 1971) which contained phenolic constituents (Brown, 1961) and soluble nitrogen compounds (O'Donovan et al., 1967). It was not until techniques for identifying these alkaloids were developed that the question of which alkaloids affected acceptability could be studied.

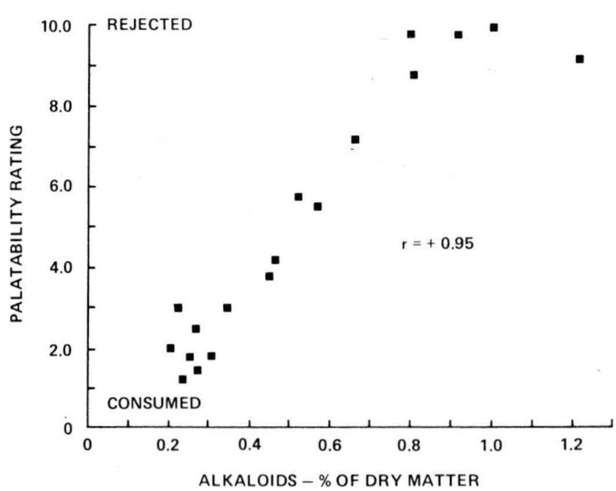

Fig. 3.4. Relationship between palatability rating and alkaloid concentration of 18 reed canary grass clones. (Source: Simons and Marten, 1971.)

It was first suggested that acceptable strains had gramine but not N,N dimethyl tryptamine (Barnes et al., 1969). Later it was observed that unacceptable strains had 18 times more 5-methoxy-N,N dimethyl tryptamine than acceptable strains; also, acceptable strains had only 0.26% total alkaloids compared with 0.78% in unacceptable strains (Barnes et al., 1969). Simons and Marten (1971) and Marten et al. (1976) substantiated that acceptability was not related to concentrations of any particular alkaloid but to the total level of alkaloids, and that this (Fig. 3.4) relationship held over a wide range of environments.

In all this work, acceptability was tested with sheep. Woods and Clarke (1974), however, found no correlation between total alkaloid levels and acceptability to cattle.

Essential oils

Many species that are important food plants for browsing ruminants contain essential oils. In California, Oh and his colleagues have been studying these in relation to deer and sheep preferences. They have found (Oh et al., 1967) that the monoterpene hydrocarbons have little effect, the monoterpene esters a mild effect, but that the monoterpene alcohols strongly inhibit rumen activity. Whether the effects on the rumen are related to palatability differences is not yet established.

Fats

There are several reports of situations where preferences of ruminants are correlated with the ether extract content of plants on offer. The most striking is that of Louw et al. (1967) who found that sheep preferences amongst species of bush in the Karoo of South Africa were related to their ether extract content, decreasing with increasing fat content. Unfortunately, there has been no attempt to identify which fats, if any, in their complex extracts influence acceptability.

Other compounds

Two striking omissions from the field studies of plant chemical compounds and food preferences have been the free amino acids and most aliphatic volatile compounds. Some forty volatile flavour substances have been identified from *Trifolium pratense* L., *T. hybridum* L., and *T. ripens* L. (Honkanen et al., 1969) but nothing is known about the response by ruminants to their odour.

Integrated responses

These discussions have dealt with single classes of compounds but it has been stated that the animal responds to the integrated effects of the stimuli from various classes of compound. This total response has not been studied to any extent. Cowlishaw and Alder (1960) and Jones and Barnes (1967)

TABLE 3.4

Weights for each chemical constituent in the vectors of principal components contributin

Principal component	Vectors of principal components			
	2	4	6	8
Chemical constituent				
Water soluble dry matter	0.32	0.03	−0.17	−0.60
Total ash	0.50	0.01	−0.41	−0.06
Total nitrogen	−0.06	0.60	0.43	−0.04
Lignin	−0.32	0.27	−0.35	−0.05
Total carbohydrates	−0.17	0.68	−0.32	−0.18
Water soluble ash	0.70	0.29	0.13	0.17
Water soluble nitrogen	−0.08	−0.10	−0.57	0.02
Water soluble carbohydrates	0.12	0.14	−0.21	0.76
Shikimic acid				
Quinic acid				
Succinic acid				
Malic acid				
Citric acid				
Total acids				
t values	2.50	2.10	4.18	−1.79
R^2		74.0		
Principal component description	Ash content	Total nitrogen and carbohydrate content	Nitrogen composition	Contrast between water soluble dry matter and water soluble carbohydrate

used the same data on preferences by sheep. The first authors looked at relationships with soluble ash, soluble nitrogen and soluble carbohydrates: the second authors looked at relationships with organic acids. Both obtained significant correlations between preference ranking and certain compounds. However, none of the compounds individually accounted for more than 64% of the variation in preferences. A ‘principal components’ analysis was made on the 14 characteristics of the herbage measured and the principal components then used as independent variables in a multiple regression analysis. This showed that 87% of the variation in preference could be accounted for by combinations of the compounds (Table 3.4). The higher the concentration of all organic acids, the higher the preference of a species, but composition of the organic acids was also important, i.e. the lower the succinic acid content and higher the citric acid content, the higher the preference. The higher the soluble ash (salt) content, the lower the preference. Also, the

essions relating preference ranking of species to the chemical composition of the species

	2	3	2	3	4	10
			0.22	0.12	0.24	−0.36
			−0.09	0.30	0.45	0.00
			−0.03	0.08	−0.25	0.32
			−0.28	−0.08	−0.23	−0.40
			0.00	0.14	−0.29	0.43
			0.19	0.33	0.50	0.19
			−0.07	−0.16	0.00	−0.32
			0.12	0.09	0.06	−0.21
8	−0.23	0.31	0.35	−0.32	−0.11	0.19
3	0.03	0.17	0.47	−0.02	0.00	0.02
.0	−0.74	0.07	−0.06	−0.60	0.31	−0.11
3	0.36	0.85	−0.31	0.25	0.12	0.03
8	0.50	−0.33	0.35	0.40	−0.39	−0.44
2	−0.16	0.18	0.50	−0.16	0.08	0.05
4	2.69	−1.89	7.42	3.33	−1.72	2.11
	80.2			87.0		
l nic ent	Contrast of succinic and citric acid contents	Malic acid content	Total organic acid content	Contrast between succinic and citric acid contents	Ash content	Lignin and citric acid contents

higher the total carbohydrates and lower the lignin, the higher the preference for a species. This analysis suggests that the animals were balancing the effects of many different components to determine the relative preference for a particular species. Another feature of the analyses is that Cowlishaw and Alder could be well satisfied that preferences were related to the mineral, carbohydrate and nitrogen composition of the herbage since the regression using these components accounted for 74% of the variation in preferences. Yet with data only on organic acid 80% of the variation was accounted for. The combined analysis suggests that the cruder carbohydrate and nitrogen measurements were dummying for the organic acids in the first analysis in Table 3.4. The whole exercise indicates the care that is needed in interpreting the relationships between preferences and a limited range of chemical constituents.

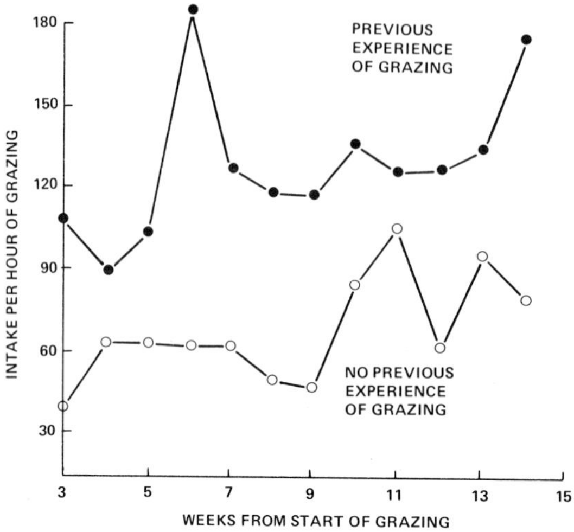

Fig. 3.5. Effect of experience of grazing on rate of intake in sheep.

TABLE 3.5

Reactions of one year old Merino wethers, differing in experience, to eight plant species (Source: Arnold, 1964b)

Species	Score for degree of utilisation [1] (scale 0—5)			
	Group 1 [1]	Group 2 [1]	Group 3 [1]	Group 4 [1]
Phalaris tuberosa	2.5	3.5	2.5	2.5
Medicago sativa	3.5	2.0	4.0	4.5
Trifolium subterraneum	3.5	4.0	3.5	3.0
Bromus mollis	1.0	1.0	1.0	1.0
Eragrostis curvula	1.0	1.0	2.5	2.5
Dactylis glomerata	2.5	2.5	4.0	3.5
Stipa hyalina	1.5	1.0	2.0	2.0
Lolium perenne	2.5	2.0	3.5	3.0

[1] Group 1. Reared at Trangie, N.S.W., on sub-humid rangelands, improved annual pastures and lucerne. The pastures differed widely in physical and chemical characteristics.
Group 2. Reared on *Phalaris*—subterranean clover pasture. The sheep experienced seasonal changes in physical and chemical characteristics of only two species.
Group 3. Reared on irrigated pastures at Deniliquin, N.S.W. Sheep experienced only green herbage.
Group 4. Reared on semi-arid rangeland at Deniliquin, N.S.W. Sheep experienced a large number of species which changed little in physical and chemical characteristics. Pasture structure was quite different from that experienced by groups 1, 2 and 3.

PREVIOUS GRAZING EXPERIENCE

Experience, particularly in early life, may alter the later behaviour of an animal, but relatively little research has been done on the effect of experience on the feeding behaviour of grazing ruminants. Lambs in a mediterranean environment learn what to select from a dry pasture by experience during the period when it is changing from a green pasture (Arnold, 1964c). If they are not allowed this experience and are transferred directly to dry feed they eat less and perform worse than lambs with experience.

The efficiency with which sheep harvest food by grazing is influenced by experience (Arnold and Maller, 1977) (Fig. 3.5). Sheep reared to 3 years of age without any grazing experience were shown to obtain much less food per hour of grazing than sheep reared on pasture.

Preferences amongst a choice of pasture plants can be strongly influenced by experience (Arnold and Maller, 1977). It was shown that sheep from rangeland areas of Australia had quite different preferences to sheep from sown pastures (Table 3.5). Differences in preferences were greater when diverse experience occurred in early life rather than in later life. It was also shown that sheep that had no experience of two generally unpalatable plant species ate much less of them than sheep with experience of similar plant material (Table 3.6). The differences in preferences were shown to persist for up to 2 years despite attempts to remove them by subjecting the sheep to a wide range of plant material.

TABLE 3.6

Differences in scores for utilisation of two unpalatable pasture species for ewes and yearlings with different grazing experience in the previous twelve months (Source: Arnold and Maller, 1977)

	Previous 12 months spent grazing			
	Phalaris pasture at Canberra A.C.T.	Irrigated pasture at Deniliquin N.S.W.	Rangeland at Deniliquin N.S.W.	Rangeland at Trangie N.S.W.
Yearlings				
Stipa hyalina	4.2	4.2	5.0	6.4
Eragrostis curvula	2.4	4.8	4.6	4.8
Ewes				
Stipa hyalina	1.0	4.0	1.0	6.0
Eragrostis curvula	4.6	4.0	5.4	6.0

AGE AND BREED AND PHYSIOLOGICAL STATE

Generally, only small differences have been found in the diets selected by sheep of the same breed but differing in age. Hodge and Doyle (1967) compared 10 week old lambs and yearlings on two types of pasture and found that they selected similar amounts of grass and clover. Arnold, Campbell and Pahl (unpublished) obtained only small differences between sheep of <12 months to >72 months of age. They found that at 5 months of age sheep had a diet that was higher in digestibility and in nitrogen content and lower in fibre than that of older sheep. This may have been due either to lambs being more selective on the individual plant or to their having a smaller jaw.

Small differences between breeds in diet nitrogen content were found by Langlands (1969). Arnold, Campbell and Pahl found that sheep of four breeds had differences in preferences over a wide range of pasture species (Table 3.7). The taste preferences of sheep of these four breeds differ (Arnold and Hill, 1972) and this may account for the differences in preferences when grazing. In cattle, no differences in preferences were found between Hereford and Santa Gertrudis cows grazing on a New Mexico range (Herbel and Nelson, 1966).

It might be thought that food preferences would change when the physiological status of the animal causes different food requirements, but there is no evidence that this occurs. The topic has been studied only on sown pastures in one environment. Both sheep and cattle selected the same diet whether pregnant, lactating or dry (Arnold, 1966c).

TABLE 3.7

Effect of breed on degree of use (0—5 scale) of eight pasture species in a grazing 'cafetaria' (Source: Arnold, Campbell and Pahl, unpublished)

Pasture species	Breed of sheep			
	Dorset Horn	Border Leicester × Merino	Corriedale	Merino
Lolium perenne	2.3	1.8	2.3	2.5
Stipa hyalina	0.5	1.0	0.0	1.8
Dactylis glomerata	0.5	1.5	1.8	2.8
Eragrostis curvula	0.0	0.0	0.0	0.0
Festuca arundinacea	1.8	1.8	2.3	2.0
Trifolium pratense	1.8	2.3	2.3	1.8
Medicago sativa	2.8	2.3	2.0	2.0
Phalaris tuberosa	2.8	2.5	2.0	2.8
S.D.	1.1	0.8	1.0	0.9

Nearly all the comparisons that have been made are between sheep and cattle. However, the earliest published comparison of the diets of different species was by Cory (1927), who included goats. The major difference was between sheep and cattle on the one hand, and goats on the other (Table 3.8) because of the amount of browsing done by goats. Wilson et al. (1975) found the same sort of differences between sheep and goats on a semi-arid woodland range in New South Wales.

On two types of mountain range in Utah, cattle ate less grass and more forbs and browse than sheep (Cook et al., 1967). On annual range in California in summer, Van Dyne and Heady (1965a) found that there were always differences between sheep and cattle but that these changed with time. For example, sheep ate more forbs than cattle in early and mid-summer but in late summer the situation was reversed. Bedell (1968) compared the two species when grazing ryegrass—clover and fescue—clover pastures (Fig. 3.6). Sheep ate at least three times as much clover as did cattle throughout the summer on the ryegrass pasture, but only did so for part of the summer when grazing fescue pasture; a difference that occurred in years of abundant feed. In a dry year, with little clover, sheep ate more of the annual grass than did the cattle.

Wilson (1976) found that cattle ate less of the short medics, certain grasses, and more dry herbage than sheep when grazing on a semi-arid grassland in New South Wales. In all these studies, and in that of Dudzinski and Arnold (1973), cattle diets were lower in nitrogen content and usually higher in fibre content than sheep diets.

The lack of consistency in differences in selection between sheep and cattle is not surprising since the physical and chemical characteristics of the vegetation being grazed change with time. Thus, whilst differences in the

TABLE 3.8

Comparison of grazing preferences of sheep, cattle and goats on rangeland in Texas (Source: Cory, 1927)

Type of feed	Percentage		
	Cattle	Sheep	Goats
Early mesquite grass	59.3	42.0	19.2
Weeds	6.9	18.9	7.9
Other grasses	8.3	7.6	3.0
Winter annual grasses	1.4	11.4	8.0
Browse	8.5	10.1	53.1
Miscellaneous	13.8	7.4	5.9

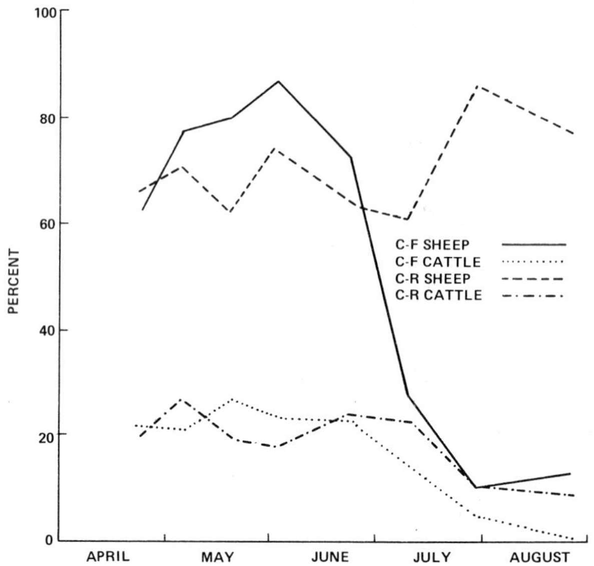

Fig. 3.6. Diet eaten by sheep and cattle when grazing tall fescue—subterranean clover (C—F) and perennial ryegrass—subterranean clover pastures (C—R). (Source: Bedell, 1968.)

mechanics of grazing will almost automatically result in different diets being consumed, the nature of the differences will be influenced by the pasture characteristics.

The relationships between pasture characteristics and differences in diets were studied by Dudzinski and Arnold (1973) on pastures varying widely in yield and botanical composition. They found, using a principle components analysis, that the pasture was described by four principal components representing (1) degree of greenness; (2) useful bulk; (3) total bulk; (4) ratio of dry grass to dry clover. The differences in most of the diet botanical components and in diet nitrogen content were related to these components. Favourable conditions, i.e. abundant green pasture, allow sheep and cattle to select the same diet. For example in Fig. 3.7 sheep and cattle ate the same proportion of clover when (a) the pasture was mainly green, (b) the pasture had little dry rubbish in it and (c) clover was dry but abundant. In Fig. 3.7 sheep and cattle obtain the same diet under the pasture conditions where the various lines cut the x axes.

INDIVIDUALITY

The diets selected by individuals in a herd or flock vary considerably in both botanical and chemical composition. The variation between individuals

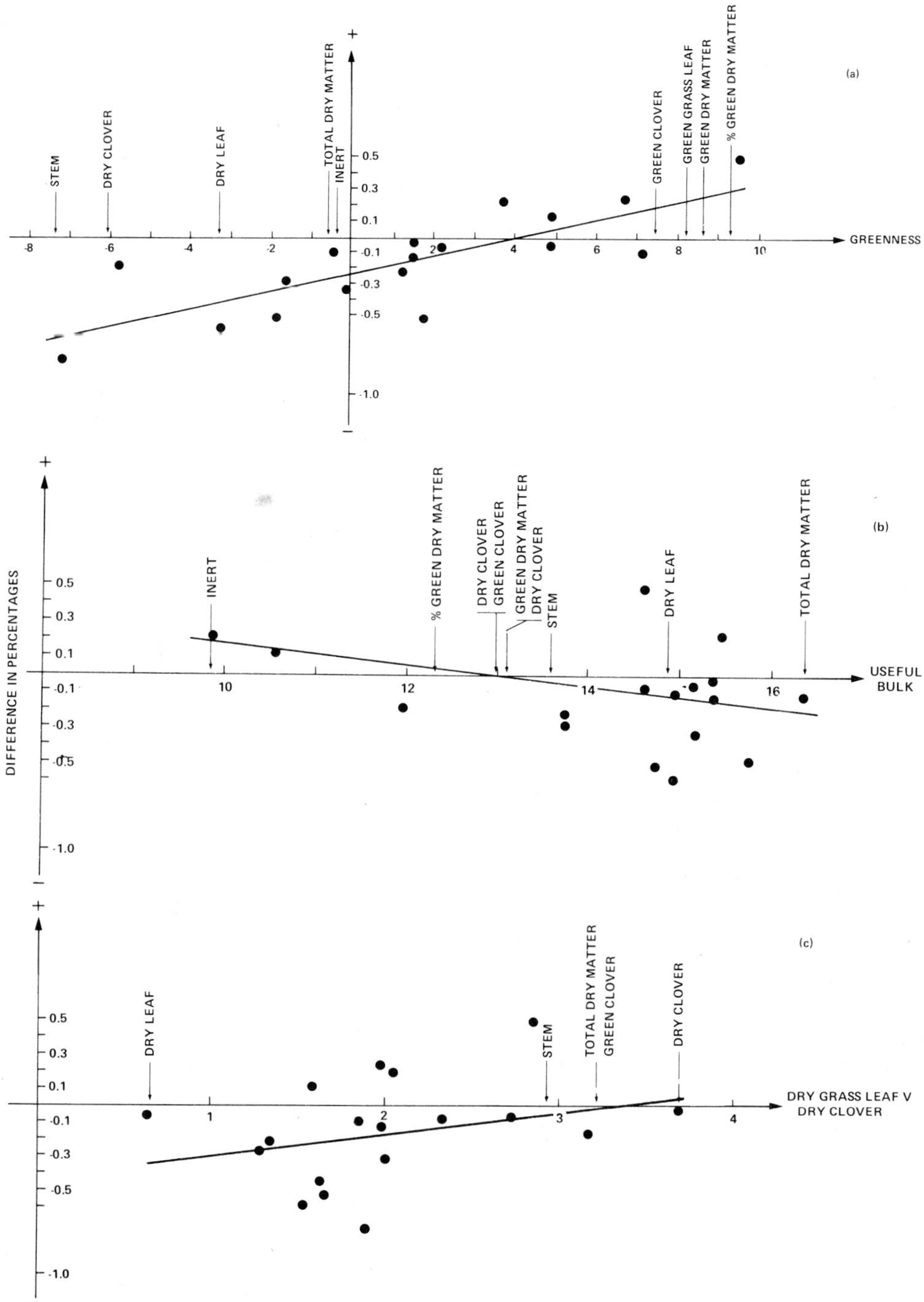

Fig. 3.7. Relationships between differences in the diets of cattle and sheep in clover content and three vectors from a principal components analysis representing the pasture they were grazing. (Source: Dudzinski and Arnold, 1973.) (a) Vector representing 'greenness' of the pasture; (b) vector representing the 'useful bulk' of the pasture; (c) vector representing the contrast between the dry grass leaf and dry clover content of the pasture. *Note*: There were 19 different pasture situations denoted as dots.

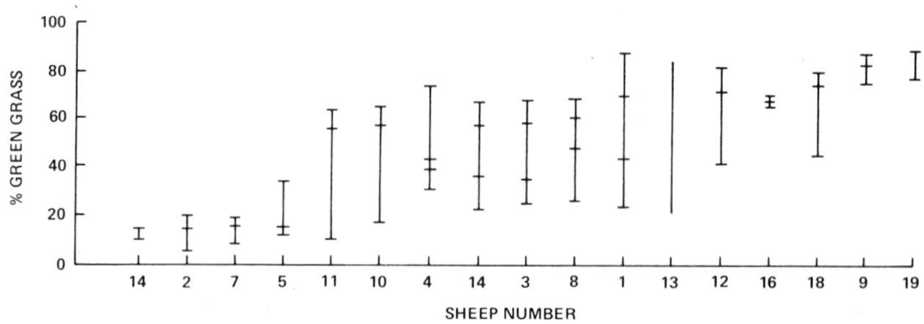

Fig. 3.8. Variation between sheep in a flock in the grass contents of their diets. (Source: Arnold, 1964a.) *Note*: The horizontal bars represent the values on different days.

is of the same order as the variation between groups of sheep from day to day. However, variation between individuals is often consistent, as is shown in Fig. 3.8. Over a week the mean grass content for the 20 sheep studied ranged from 10 to 70%. This sort of variation means that a herd or flock may not be very selective as was shown earlier, even though the individuals are highly selective. The overall degree of selectivity depends on the range of acceptability in the components of the rangeland or pasture being grazed.

DIURNAL VARIATION IN THE DIET

Both sheep and cattle change their diets during the day. This occurs in part because of changes in location of grazing during the day, particularly on rangeland. A question of interest is whether the diurnal changes are consistent and if so do they reflect a feedback or change in diet due to degree of satiation? The evidence is conflicting.

Arnold et al. (1964) found considerable within-day variation in the diets of sheep grazing sown pastures but no consistency, whilst Langlands (1965, 1967) found consistent changes in diet nitrogen content (Fig. 3.9) but not in diet digestibility. Langlands' data show an increase in diet nitrogen content as the day progresses — because the animals become more selective? Sheep grazing on desert rangeland had an increased content of nitrogen and gross energy in the afternoon (Kothmann, 1966). Cattle, but not sheep, had a higher diet nitrogen content in the afternoon when grazing annual range in summer in California (Van Dyne and Heady, 1965a).

Obioha et al. (1970) found that cattle diets varied within the day, but not consistently in chemical composition. They did find, however, that in the morning more grass than forbs was eaten and in the evening the reverse. This suggests that, because the grass was more accessible, selection was influenced by a desire to eat quickly during the morning grazing period. This is supported by Arnold et al. (1964) who showed that diets of sheep contained more dry material and less nitrogen when the sheep had been fasted over-

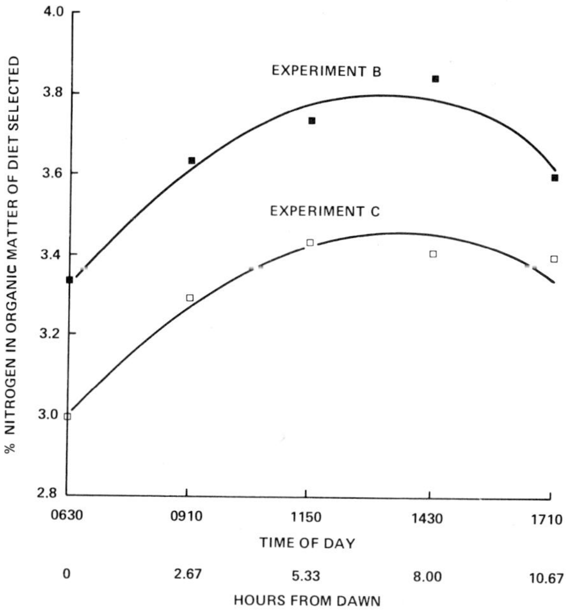

Fig. 3.9. Changes in the nitrogen content of the diet of sheep during the day in two experiments. (Source: Langlands, 1965.)

night. Also, Van Dyne and Heady (1965b) found that sheep and cattle grazed more easily accessible grass and forb heads in the morning, but more leaves, which are less accessible, in the afternoon (Fig. 3.10).

NUTRITIONAL WISDOM

Just how wisely the ruminant is able to select its food is often debated. Few, if any, critical experiments have been made from which this question can be answered. Whilst the persistence of a species must indicate that, on balance, the sensory responses have been developed to give adequate nutrition, the ruminant, like all other animals, is hedyphagic (i.e. selects food to minimize unpleasant and maximise pleasant, or both, olfactory and other sensations — McClymont, 1967) and may often select plants not for their nutritional advantages but for their flavour. Domestic species are undoubtedly subjected by man's stock feeding techniques to sensory stimuli outside their inherent sensitivity ranges, which could explain why sheep, but not kangaroos, in Western Australia preferentially eat plants of *Gastrolobium* and *Oxylobium* species, whose monofluoracetate content is lethal. It may thus be unreasonable to ask "do domestic ruminants show specific euphagia, i.e. preferentially select plants to meet specific nutrient needs" because the food

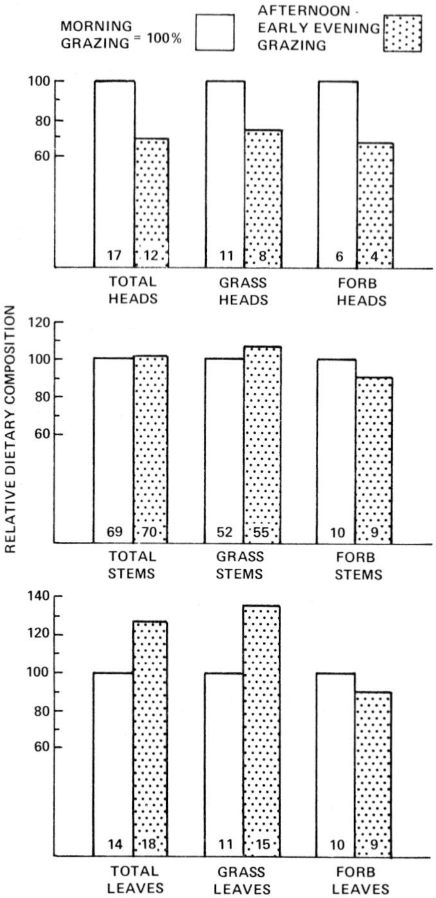

Fig. 3.10. Differences in the composition of diet of cattle and sheep in the morning and afternoon grazing a dry annual rangeland in California. (Source: Van Dyne and Heady, 1965a.) *Note*: Proportions in the morning diets are shown as 100%. Actual percentages in the diet are given at the base of each bar.

sources available may be alien in that they did not evolve with the animal species as in a natural ecosystem.

Within a single plant species, sheep may select plants with the highest phosphate content (Reid and Jung, 1965; Ozanne and Howes, 1971), when the animal is not suffering from phosphate deficiency. By contrast, Gordon et al. (1954) found that grazing sheep and cattle failed to correct a phosphate deficiency when given the opportunity. Gordon and Tribe (1951) showed that sheep did not select from basic milled foodstuffs a diet adequate for bearing and rearing a lamb. On the other hand, sodium deficiency in housed sheep was rectified very precisely when sodium solutions were made available (Denton and Sabine, 1963). When grazing, sodium deficient sheep may preferentially select herbage species high in sodium (Arnold, 1964a), but in a detailed study (Arnold, unpublished) it appears that they cannot detect levels of less than 0.10% in herbage and do not select for sodium below this level even though it is more critical to do so than when levels are higher.

Thus care must be taken in judging if there is any motivation in selective grazing. It would seem that innate reactions mainly determine selection by grazing animals, and that these reactions, euphagia, have developed during the evolution of each species. However, euphagia is not necessary for survival under husbandry conditions, as shown by Dove (1935) in experiments with chickens.

The selection of green herbage in preference to dry, which always occurs, is nutritionally advantageous in many situations because nutrient content and digestibility are usually higher in green herbage. On luxuriant improved pastures, selection of leaf tips and young leaves results in a diet richer in protein and lower in readily fermantable carbohydrates than is nutritionally desirable (Arnold et al., 1966). Sheep on dry herbage may not select the legume component, at a time when this could be nutritionally advantageous, because it is in the least accessible position in the plant canopy (Arnold, 1964d). Thus if we assume innate euphagia, there is much evidence that this may be secondary to hedyphagia. Also, innate euphagia is not manifest when the mechanics of grazing result in the removal of the top canopy of feed first, even though the highest quality feed is at the base of the pasture.

PREFERENCES AND FOOD INTAKE

Grazing of any plant community will always result in the preferential selection of certain species, at least during some part of the year. The animals will often continue to graze on preferred species even when availability is poor. There is a resultant reduction in total intake and in productivity. It is open to question whether the neglected species would be eaten if the preferred species were absent. This is a critical question and yet there is very little available information. Experimentally, it is difficult to test since other factors such as herbage yield, pasture structure and digestibility of the diet must be constant to all species under test.

Table 3.9 gives the food intakes of Merino sheep grazing seven pastures with yields high enough not to limit food intake. There were considerable differences in digestible organic matter intakes from pastures producing the same diet digestibility, e.g. ryegrass, cocksfoot, phalaris and lucerne. These differences reflect differences in acceptability. The ranking of species in intake does not agree with that from a free choice amongst the species.

Comparisons of acceptable and unacceptable strains of *Phalaris arun-*

TABLE 3.9

Reaction of one year old Merino wethers to different pasture species (Source: Arnold, 1964b)

Pasture species	Green herbage per unit area (kg ha^{-1})	Relative daily digestible organic matter intake	Relative preference ranking [1]
Lolium perenne	1666	100	79
Dactylis glomerata	1469	89	89
Phalaris tuberosa	1875	79	79
P. tuberosa + Trifolium suberraneum	3771	85	—
T. subterraneum	2978	60	100
Medicago sativa	2801	79	100
Native pasture	586	73	—

[1] Preference ranking test based on free choice of eight species which included five of the species used — test conducted at the same time as the above intake test.

TABLE 3.10

Intakes of digestible organic matter by sheep with senses impaired relative to those of sheep with all senses intact (Source: Arnold, 1966b)

Species	Senses impaired			
	None	Smell	Taste	Touch
P. tuberosa	100^{c}	84^{ab}	91^{bc}	78^{a}
D. glomerata	100^{b}	65^{a}	123^{c}	118^{c}
M. sativa	100^{b}	108^{b}	80^{a}	103^{b}
T. repens	100^{a}	94^{a}	118^{b}	91^{a}
E. curvula	100^{b}	86^{ab}	161^{c}	78^{a}

Species	Senses impaired				
	All	Smell	Taste	Touch	None
P. tuberosa	100^{c}	73^{a}	105^{a}	89^{b}	72^{a}
D. glomerata	100^{c}	68^{ab}	80^{b}	75^{b}	59^{a}
M. sativa	100^{b}	116^{b}	111^{b}	71^{a}	86^{ab}
T. repens	100^{ab}	111^{ab}	83^{ab}	81^{a}	117^{b}
E. curvula	100^{ab}	104^{b}	100^{ab}	127^{c}	87^{c}

Figures with no letter in common are significantly different at the 5% level of probability on a line.

dinacea have produced differences in intake up to 36% in favour of acceptable strains (Roe and Mottershead, 1962; O'Donovan et al., 1967; Marten et al., 1976). These results do not show why intakes differed, but the authors imply that odour is of importance. Arnold (1966b) found that on five of 11 pastures, intake was influenced either by taste, smell or touch with decreases in intake of up to 61% and increases of up to 35% due to sensory stimuli (Table 3.10). Thus evidence is accumulating that the acceptability of pasture plants can strongly influence the food intake of grazing animals.

Previous experience can influence intake; effects are usually small, but intake of sheep inexperienced on a pasture may be lower by 50% than that of experienced sheep (Arnold and Maller, 1977). The time needed for adaptation is not known.

PLANT COMMUNITY, INTAKE AND GRAZING BEHAVIOUR

The yield and physical presentation of feed available to the grazing animal may have marked effects on feed intake under intensive pasture conditions (Broster et al., 1963; Arnold, 1964d; Greenhalgh et al., 1966; Arnold and Dudzinski, 1967a, b), but may have no measurable effect under other, particularly extensive, conditions. This range of effects is to be expected because of selective grazing. The problem in studying relationships between amount of food available and food intake is to know from what elements of the plant community the animal makes its selection so that the relevant pasture variables can be measured.

Even for pastures of a single plant species there is rarely, over a short time span, a simple relationship between intake and pasture yield (e.g. Wheeler et al., 1963). The extent to which intake is kept below the level determined by energy demand and the chemical and physical attributes of the diet, depends on the adaptability of grazing behaviour. A simplified model of intake, grazing behaviour and pasture condition was given by Arnold (1964a) and elaborated on by Arnold and Dudzinski (1966).

Where availability changes rapidly either during the day or from day to day, e.g. with strip grazing by cows, the amount allocated per animal per day must influence intake (Broster et al., 1963; Greenhalgh et al., 1966). With a more extensive system, Hodgson and Wilkinson (1968) found that intake of cows, heifers and calves was unaffected when availability was reduced from 8000 to 2000 kg ha^{-1}.

With sheep grazing green pasture, the digestible organic matter intake of lambs (Arnold, 1964e), year old wethers (Arnold and Dudzinski, 1967b), and ewes of four breeds in various reproductive states (Arnold and Dudzinski, 1967a; Arnold, 1975), was shown to be related to pasture condition. Intake was related to yields per unit area of green and dry pasture, length of pasture, numbers of leaves per unit area and diet digestibility. In multiple regression models, between 20 and 80% of the variability in intakes was accounted for. Other unmeasured variables were important, but it is not obvious what they were. The known variables influencing intake changed with different types of green pasture; however, they accounted for little of the variability in intake when the pastures were approaching maturity or when they were dry. We have here another multivariate situation yet to be explained.

Homeostasis of intake with changing pasture conditions is maintained by altering grazing time, bites per minute and amount per bite. There is no set pattern of adjustment to meet a particular energy demand under different pasture conditions (Arnold, 1970), although specific relationships have been obtained between these variables and pasture yield in specific situations (Arnold, 1960a; Arnold and Dudzinski, 1966; Leigh et al., 1968). The interrelationships between the operative variables have not been fully explored, but it is of interest that sheep with different energy demands (due to age, size or reproductive state) maintain the same differences in intake over a wide range of pasture conditions (Arnold, 1975).

The social use of space, social hierarchy and social facilitation may influence intake of grazing animals but there are very few quantitative data. Under extensive grazing, defined and limited home ranges may exist, producing different nutritional regimes for different animals (Hunter and Milner, 1963). Grazing use of an environment is restricted by slope, distance to water, etc. (Gonzales, 1964; Cook, 1966), and intake of the animals will be restricted at times as a consequence.

Chapter 4

SEXUAL BEHAVIOUR

BEHAVIOUR IN FERAL AND WILD POPULATIONS

In wild and feral sheep and goats, the sexes naturally segregate for much of the year, with most of the males forming batchelor flocks or herds. Sexual activity between the sexes is strictly seasonal in the period known as the rut. The pattern of behaviour is similar in all three species and has been described in sheep by Spencer (1943), Geist (1969) and Grubb and Jewell (1973), and in goats by Crook (1969), Schaller and Laurie (1974) and McDougall (1975).

A few weeks before the females start oestrus cycles the bachelor herds break up and the males join the groups of females on their home ranges, moving from one group to another. A detailed description of the pattern of behaviour in feral Soay sheep has been given by Grubb and Jewell (1973) and by Grubb (1974). Strange rams begin appearing in an area and begin investigating ewes. This develops from casual contacts into systematic investigations of ewes involving nosing the vulva, nudging the ewe, and examining her urine. During the rut such investigations of many ewes in succession become frequent and occur at any time of the day. When a ewe coming into oestrus is discovered she is chased by parties of males, including ram lambs. The chase is usually led by a large ram, running in a low stretch posture, and giving brisk turns of the head to one side or the other. Often he will round on any followers and nudge them.

At the same time that investigations of ewes intensify, so confrontations between males involving agonistic displays and fighting become more frequent. The peak of agonistic activity, as measured by the frequency of nudging displays, precedes the peak in overt oestrus in the ewe population (Fig. 4.1).

Rams of highest dominance rank achieve many more tending bonds than those lower on the scale; these latter animals establish a high proportion of their bonds later in oestrus when more powerful rams have left the ewes.

During the rut the grazing activity of all males is seriously disrupted as shown in Fig. 4.2. The consequence is reduced nutrition, and the entering into winter with low body reserves and consequent high mortality in the young males which at that time have least tissue reserves. The involvement of the young rams must be part of their learning process; the ontogeny of sexual behaviour in Soay rams is shown in Table 4.1.

In horses, harems are formed by dominant males and kept throughout the

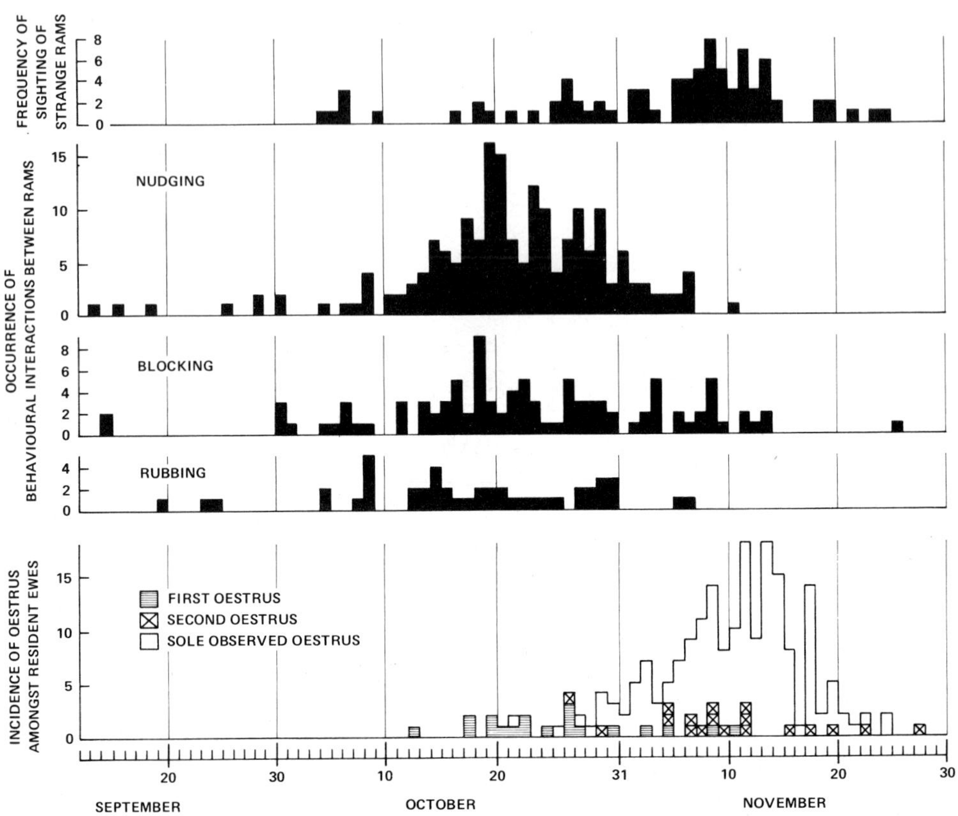

Fig. 4.1. Frequency of sighting of strange rams in a ewe home range, and of the occurrence of behavioural interactions between rams, set in their time relations to the incidence of oestrus amongst the ewes. (Source: Grubb and Jewell, 1973.)

TABLE 4.1

Ontogenic differentiation of behaviour patterns involving nudging and mounting (Source: Grubb, 1974)

(1)	Ram lambs (first two months): nudging display incomplete (no empty licking, rutting call, penis erection) and in essentially a single context; mounting in a 'non-sexual' context.
(2)	Ram lambs (next six months): nudging display now complete (rutting call, empty licking, twist, kick; and penis erection?); mounting of adult ewes may occur.
(3)	Ram lambs (first rutting season): nudging now shown in two distinct contexts with different associated behaviour; mounting of oestrous ewes shown in definitely sexual context; mounting of clashing or blocking males also occurs.
(4)	Younger rams (1.5—4.5 years old): the same, but mounting almost completely confined to sexual contexts; mounting of other rams may occur, but rare.
(5)	Older rams: the same, but nudging many form part of a 'mixed' display, in sexual contexts only; mounting of other rams never shown apparently.

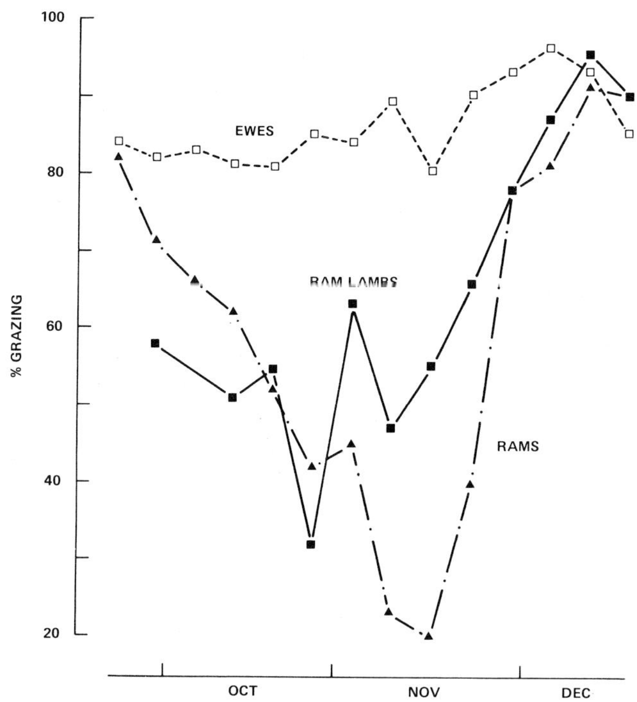

Fig. 4.2. Grazing activity of Soay ewes, ram lambs and older males during the rut. (Source: Grubb, 1974.)

year, as described in Chapter 2. Numerous interchanges of mares between harems occur (Hall, 1972) due to challenges from stallions with no harem. Camels also form harems (Gauthier-Pilters, 1974).

GENERAL PATTERN OF SEXUAL BEHAVIOUR IN DOMESTIC POPULATIONS

Man probably has more influence on the sexual behaviour of his animals than he does on any other type of behaviour. The sexes are usually separated for much of the year and only sexually mature animals are brought together for mating. Thus development of sexual behaviour in the male is likely to be totally different from that described by Grubb (1974) in feral sheep. The ratio of males to females is also controlled. The concern here will be on the broad patterns of behaviour of males and females leading up to mating and the effects of man's controls on reproductive efficiency in free-ranging animals. The patterns of behaviour in courtship and mating are the same in pens or under free range, and will not be described. Detailed descriptions and reviews are available elsewhere (e.g., Fraser, 1968; Blockey, 1976).

There is seasonal breeding in most of the animal species being considered here, with the natural stimulus to bring females into oestrus being shortening daylength (Hafez, 1959). There is a marked breed difference in sensitivity to daylength and there may be a requirement for some absolute quantity of darkness.

The induction of oestrus is not only due to endogenous changes in the female animal but also to exogenous stimulation by odour, sound, sight or touch supplied by the association with an active male. It has been found that the proportion of ewes in oestrus is affected by the period of contact with rams (Riches and Watson, 1954; Lamond, 1963). The assumption is that the odour of the ram acts as a pheromone: this is substantiated by work of Morgan et al. (1972a) who found that fewer than half of a flock of anosmic Border Leicester X Merino ewes became pregnant whereas nearly all those with the sense of smell did.

Under very hot conditions oestrus will be suppressed even in breeds adapted to an environment. At Julia Creek in Queensland the provision of shade significantly increased the proportion of ewes in oestrus when maximum temperatures were around 35° C (Hopkins and Pratt, 1976).

A seasonal variation in sexual activity exists in males and has been described specifically in rams (e.g., in Targhee rams in California — Pepelko and Clegg, 1965 and in Romney and Cheviot rams in New Zealand — Shackell et al., 1977). Border Leicester rams were found (Lindsay and Ellsmore, 1968) to have a more seasonal pattern than Merino and Dorset Horn rams, being more active sexually in autumn and spring (Fig. 4.3), but whether the

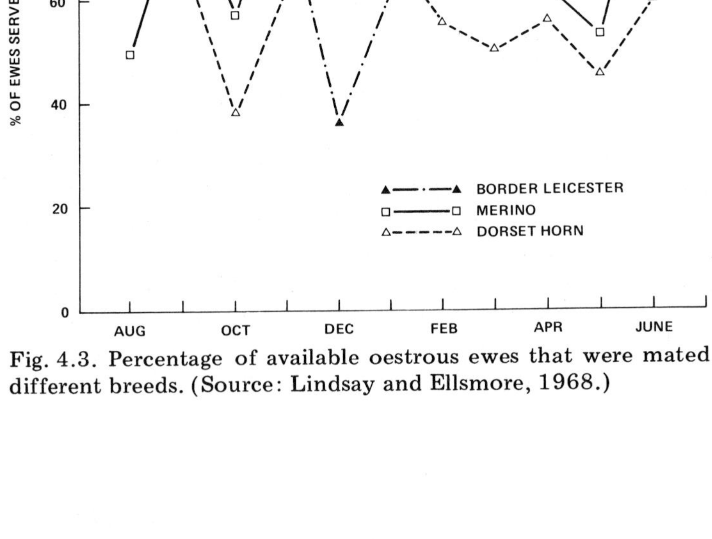

Fig. 4.3. Percentage of available oestrous ewes that were mated each month by rams of different breeds. (Source: Lindsay and Ellsmore, 1968.)

difference is due to daylength or temperature is not known. There is some indication that libido in rams is influenced by daylength (Moule, 1950). Libido is reduced at high temperatures in both rams and bulls but this is not due to endocrine changes, just to disinterest due to discomfort according to Yeates (1965).

BEHAVIOUR OF THE MALE

Male sex drive is stimulated by the odour of the oestrus female. In some species, such as the horse, the odour is spread deliberately through frequent urination. Without the sense of smell rams find it harder to pick ewes in oestrus from those that are in anoestrus (Lindsay, 1965). Van Sambraus and Waring (1975) found that bulls respond to the smell of the urine of a cow in oestrus but that they receive other specific stimuli as well. These may be auditory, as suggested by De Vuyst et al. (1964), and visual (Hale, 1966). Males of all species will spend a good deal of time searching for females in oestrus. Robertson and Rakha (1965) found rams doing this throughout the day and night.

As will be seen later, spatial distribution of males and females is important and detection is poor in certain situations which suggests that the odour of a female in oestrus can be detected only over a relatively short distance and other cues from the behaviour of the female and of other males are vitally important.

There is no easy way to assess libido under controlled conditions, e.g. with oestrus females in yards, and relate it to mating performance in the field. Although some correlation has been found for young Merino rams between libido score measured in pens and number of services and number of ewes that lambed after field mating (Mattner et al., 1971, 1973), other workers have found no correlation at all (Hulet et al., 1964; Cahill et al., 1975; Kelly et al., 1975; Walkley and Barber, 1976).

The most important factors determining pregnancy rate are numbers of viable sperm received by the female. In ewes the number of pregnancies in a flock is related to the number of services (Fowler, 1975). This will be influenced by the ratio of males to oestrus females and by other factors as discussed later. It will also be influenced by dominance ranking in the males, particularly if a dominant animal is infertile.

Inkster (1957) noted that some rams were superior to others in persuading ewes to stand for service and that this was not due to the vigour of his teasing but to some other factors not obvious to the observer, but readily apparent to the ewe. This characteristic may be associated with social dominance although there is no evidence one way or the other. However, dominant Merino rams have been shown to give a higher lambing percentage than subordinate rams when mated singly (Marincowitz et al., 1966).

Dominance amongst males during mating has been studied only in rams. Aggressive acts were more frequent when two Suffolk rams were together than when there were three (Shreffler and Hohenboken, 1974); they also increased as the number of ewes in oestrus decreased. The dominant ram did not have the highest scores for all attributes of mating, e.g. percentage of ewes mated, number of mounts per oestrous ewe or total number of mounts. Dominance ranking in the rams was reasonably linear and stable.

It was observed by Arnold and Baas Becking (1964) that the dominant ram in a group of four was the least active in seeking for ewes in oestrus and not the most aggressive. However, if he wanted a particular ewe he took her, and he was rarely challenged. Challenges went on between lower ranking rams.

Bourke (1967) reported a similar finding. In a group of three rams, those ranked 2 and 3 moved around the paddock sniffing ewes' genitals. If the dominant ram observed either of these rams mounting a ewe he would run up to 300 m to challenge and then lead the ewe off to where he had been with other ewes in oestrus. Unlike Arnold and Baas Becking, Bourke observed no physical fights, only 'mental' conflicts involving the approach of the dominant ram with head lowered and the retreat of the subordinate ram before physical contact. During 17 days the dominant ram mated with 124 ewes whereas the other two rams mated with only 61 and 54 ewes. During these days there had been 153 ewes in oestrus. Thus the dominant ram was the most successful. On the day with the highest number of oestrous ewes (17) he mated with 15 of them. Much the same story has been found with Romney rams, dominant rams providing the majority of services but not being the most aggressive animals (Lambourne, 1956).

Thus there is considerable variation in the proportion of ewes a ram may mate. In flocks with four rams the range found can be from 4 to 96% (Knight and Lindsay, 1973). In terms of number of services a day, Mattner et al. (1967) observed a range of 8—38. This variation will be partly due to dominance effects and partly to differences in libido.

BEHAVIOUR OF THE FEMALE

The behaviour of the female in oestrus varies between species and is summarised in Table 4.2. Oestrous cows will seek bulls (De Alba and Asdell, 1946; Wohanka, 1962). Ewes will migrate to rams if these are tethered (Lindsay and Robinson, 1961b; Lindsay, 1966; Lees and Weatherhead, 1970; Allison and Davis, 1976b). Once with a ram a ewe will not seek another ram (Lindsay, 1966). The length of stay may be long — Bourke (1967) observed one stay of 36 h — or short, much depending on the ram. If the ram has the choice of many oestrous ewes, or is aggressively dominant, it may well move off rapidly to prevent another ram mating. The ewe will not follow a ram in such a move.

TABLE 4.2

Brief characteristics of heat in some ungulates (Source: Fraser, 1968)

Animal	Typical oestrous behaviour
Horse	Urinating stance repeatedly assumed; tail frequently erected; urine spilled in small amounts; clitoris exposed by prolonged rhythmical contractions of vulva; relaxation of lips of vulva. Company of other horses sought; turns hindquarters to stallion and stands stationary.
Cow	Restless behaviour; raises and twitches tail; arches back and stretches; roams bellowing; mounts or stands to be mounted; vulva sniffed by other cows.
Sheep	During pro-oestrus there may be a short period of restlessness, courting ram with accepting mating. In oestrus proper ewe seeks out ram and closely associates with it; may withdraw from flock. Remains with ram when flock 'driven'. Tail wagging may be shown rarely.
Goat	Restless in pro-oestrus. In heat most striking behaviour includes repeated bleating and vigorous, rapid tail waving; poor appetite.
Reindeer	Maintains bond with bull; separates from herd.
Bactrian camel	Crosses neck with male; adopts sitting posture with male.

Competition between ewes for the attention of a ram have been suggested to occur (Lambourne, 1956; Inkster, 1957; Lindsay and Robinson, 1961b). However, there have been no descriptions of the nature of the competition in the field. In pens, Hulet et al. (1962) observed ewes nudging one another away from a ram. Competition between old and young ewes is suggested as one cause for a lower conception rate in young ewes.

The grouping of oestrous ewes around a ram has been termed a harem and it has been suggested that this grouping is arranged deliberately by the ram. However, it seems more likely to be an unorganised occurrence. A ewe in oestrus once found by a ram will stay with it so long as the ram does not race off to challenge another ram. If it moves off to challenge, the ram may well return to where it had been before — followed by the new oestrous ewe. This 'harem' formation has been observed relatively infrequently. It is found when ram density per oestrous ewe is low (Croker and Lindsay, 1972) and where stock density is also low, probably because the rams are dispersed and there is less general disturbance from other rams.

EFFECTS OF MANAGEMENT CONTROLS

Type of paddock

The farmer or rancher has to decide the size of the paddock to be used for mating. He also decides the ratio of males to females and whether the maiden females are mated along with the older females. Both factors may affect the reproductive rate in his flock or herd because they can influence mating behaviour by affecting the frequency of contact between oestrous females and males.

There are conflicting reports on the influence of paddock size on reproductive performance. Marincowitz et al. (1966) found no difference in number of ewes mated by single rams joined with 130 Merino ewes in flat paddocks of 43 and 200 ha in South Africa. In New Zealand, Lambourne (1956) found that the proportion of mature Romney ewes lambing was the same on rough terrained large paddocks as on small flat paddocks. However, fewer maiden ewes (56%) lambed on the large paddocks than on the small paddocks (73%). More recent studies in New Zealand (Allison and Davis, 1976a) produced similar results but it was apparent that shape and topography of the paddock were more important than size (up to 274 ha) over a range of ewe : ram ratios. A feature of these investigations was that a proportion of maiden ewes failed to mate in paddocks where the flock split up into groups. In such a situation migration of oestrous ewes to rams may be of greater importance and the maiden ewes migrated less than mature ewes.

In Australia, Lindsay and Robinson (1961a, b) found detection of oestrous ewes by rams was sometimes equal on paddocks of 0.2 and 7.0 ha, but sometimes was higher on the smaller paddocks. No reason could be suggested for the variable results. However, in another comparison (Nicholls et al., 1976) the proportion of ewes served by rams was found to be similar in paddocks of 0.6 and 6.0 ha, but 3% fewer ewes lambed in the 6 ha paddocks and if the 6 ha paddocks were irregular in shape and topography, the figure was 5% fewer ewes lambing. In this case the effect is presumably on the number of services per ewe.

No detailed studies have been made with cattle, but Jardine and Anderson (1919) state that on rough rocky rangeland terrain the number of bulls needed to keep up the calf crop nearly doubles that needed on level ground. The reason, presumably, is that the bulls have greater difficulty locating oestrous cows.

Ratio of males to females

The ratio of males to females is important in these situations. Beerwinkle (1974) found that bulls mounted 100% of oestrous cows when there were 30 cows per bull, but only 64% with 60 cows per bull and 51% with 100 cows per bull. Tomkins and Bryant (1972) found that with one ram per six oestrous ewes a ewe was served 8.5 times in an oestrous period but this fell to 3.7 times with one ram to 30 oestrous ewes. Also, a ewe was being served over only one fifth of its oestrous period instead of over two fifths and conception rate decreased from 75 to 56%. Another aspect of the ram : ewe ratios is the number of rams a ewe is served by, because if any are infertile the chances of pregnancy increase with an increasing number of rams serving each ewe. With three rams per 100 Dorset Horn ewes, 15% were served by only one ram, 34% by two rams and 51% by three rams (Tomkins and Bryant, 1972), the average number of services being 4.2 per ewe from all rams.

There is virtually no information for other species on frequency of service although Sumner et al. (1968) observed bulls and rams over a 12 h daylight period and obtained figures of 1.7 services per oestrous cow compared with 4.0 services per oestrous ewe.

Age

The figure of 4.2 services per ewe obtained by Tomkins and Bryant was similar to that found by Lambourne (1956) in mature Romney ewes, but maiden ewes were served only 1.9 times, on average. In addition there was a large difference in the interval between first and last service, 3.3 h for maidens compared with 15.0 h in mature ewes. Since maiden ewes have a shorter oestrous period they will be served less often and their chances of becoming pregnant will be less when the ram : ewe ratio is small.

There have been no definitive studies to show what factors contribute to fewer maiden ewes in oestrous being detected. They may stimulate the rams less, and may seek rams less actively. Certainly rams prefer adult ewes, as Moran (1974) found that the proportion of maiden Border Leicester × Merino ewes returning to service increased as the proportion of mature ewes in the flock decreased.

Young ($1\frac{1}{2}$ year old) rams are less active sexually than older rams. For Merinos, Croker and Lindsay (1972) found that with 1.2 rams per 100 ewes, $1\frac{1}{2}$ year old rams mated 74% of the ewes whereas $2\frac{1}{2}$ year old rams mated 90% of the ewes. Lightfoot and Smith (1968) obtained very similar results measured in terms of percentage of ewes lambing. With $1\frac{1}{2}$ year old rams 79% of ewes lambed when there were two rams per 100 ewes and 91% with four rams per 100 ewes.

Intensity of management

It is under extensive ranching or pastoral conditions that contact between males and females might be expected to be less frequent than under intensive conditions. This has not been adequately studied.

The authors observed 370 ewes and six rams in a 4000 ha paddock at Broken Hill, New South Wales. During the day the flock split up into 7—16 sub-groups and, on average, only one in three had rams in the group. At least one group had two rams and the groups changed little in composition during the day. Thus any ewe in oestrus had a low chance of being served during the day. In a flock of about 320 Merino ewes grazing on a 5100 ha paddock at Meekatharra, Western Australia over three mating seasons, the percentage of ewes without contact with rams during the day increased with decreased sub-group size (Fig. 4.4) and the proportion of ewes mated was from 30 to

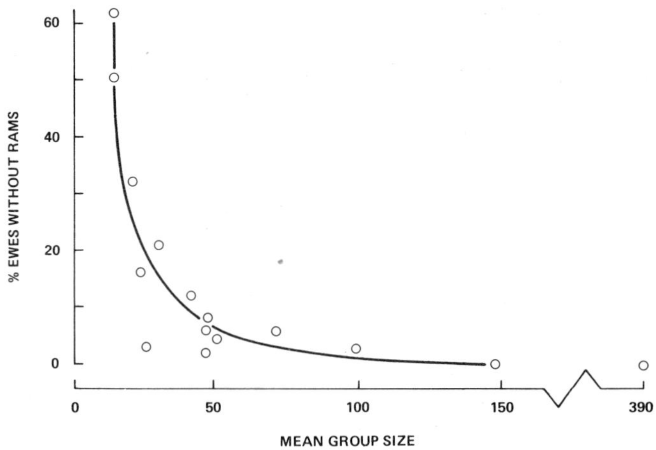

Fig. 4.4. The relationship between mean group size and the percentage of ewes without rams. (Source: Gardiner et al., 1978.)

47% less in the first 17 days of mating than for similar ewes mated in a small paddock. Further analysis of the study (Dudzinski et al., 1978) showed that rangeland conditions affected sheep dispersion and the average chance of ewes being with a ram. The average percentage of ewes in groups without rams was 2% in a good year, 6.5% in a medium year and 26% in a poor year.

Fletcher and Nicolson (1976) found that a flock of 400 Merino ewes

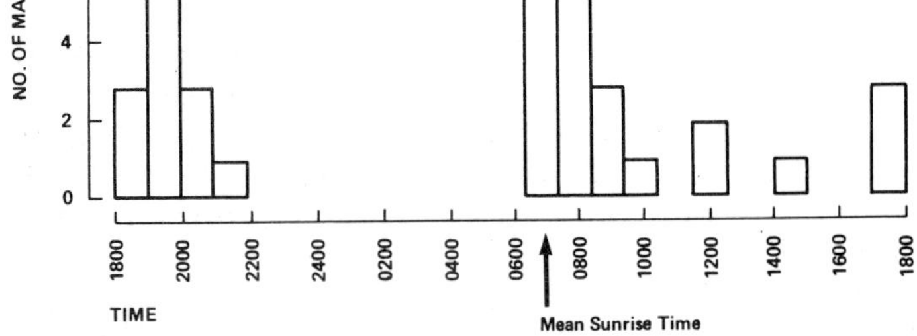

Fig. 4.5. The incidence of matings observed in a flock of Cheviot ewes. (Source: Fraser, 1968.)

grazing on 1500 ha of semi-arid woodland—shrubland formed into small groups, often with single rams, but they congregated at water in the early morning and/or evening into larger groups with several rams. The extent of aggregation in the early morning would determine the reproductive rate for two reasons: ewes in oestrus may join a group with a ram and peaks of mating occur at these times (Fig. 4.5). This has been observed in Welsh Mountain (Hutchinson et al., 1964), Cheviot and Suffolk sheep (Fraser, 1968) and in Merinos (Blockey and Cumming, 1970; Croker and Lindsay, 1972). A similar pattern exists in cattle (Rollinson, 1960; Von Sambraus, 1968). Presumably this is related to the time of day when ewes and cows come into oestrus.

Under extensive conditions where sheep do not need to aggregate to drink or to camp, contact between ewes and rams will be low. This is likely to occur when feed is scanty but surface water is available in several locations.

Interaction between breeds

Reproductive rate in flocks of ewes of one breed and rams of another breed may depend on the stimulation of the rams' sex drive by the ewes. Hafez (1951) studied ewes of six breeds mated with rams of three breeds and considered that a high sex drive in rams occurred with Dorset Horn and Suffolk ewes which actively sought the rams. Sex drive was lower with Romney and Border Leicester ewes which only approached rams if they were nearby. Sex drive was low with Blackface and Welsh ewes which made no attempt to approach rams. Inkster (1957) also noted that Romney ewes were "markedly less enthusiastic about service than Romney × Cheviot ewes".

One final aspect is the preference of rams for ewes of different breeds when such flocks are formed. Dorset Horn, Merino and Border Leicester rams were found to have no preferences amongst ewes of six breeds (Lindsay, 1966). However, Bourke (1967) found that three Merino rams run with 75 Merino and 75 Border Leicester ewes showed a preference for the Merino ewes, which averaged two services by the rams, compared with one service for the Border Leicesters. A feature here was that the two groups of ewes stayed apart in the paddock.

There is other evidence of preferences for the opposite sex of the same breed. Lees and Weatherhead (1970) found that Clun Forest ewes preferentially chose Clun Forest to Suffolk or Hampshire rams when the rams were tethered. With free movement, Arnold (unpublished) found that the first service of ewes of four breeds was usually by rams of the same breed. This breed preference relates to breed identity and social organisation as discussed in Chapter 2.

When males of different breeds are competing then breed dominance may have a significant effect. Donaldson (1962) observed that when 21 Hereford and 21 Brahman bulls were mated with 600 Hereford cows only 10% of the calves born were Brahman crossbreds. Yet when only Brahman bulls were used the calving percentage was the same as when only Hereford bulls were used. This result suggests that the Brahman bulls will not compete with Hereford bulls.

Chapter 5

MATERNAL BEHAVIOUR

Close to parturition the behaviour of the female changes and after parturition even more so. In herding and flocking species the changes in social behaviour are most marked because the desire to stay with the flock or herd is subjugated to the needs of her offspring. Normally, she will not return to the herd or flock until either the young can follow or can be safely hidden. Even in the ensuing months until weaning, the behaviour of the female is dominated by the care given to her offspring.

As Rheingold (1963) has pointed out, maternal behaviour occupies a central position in the lives of both the mother and offspring, in the social organisation of the species, and in the preservation of the species. For the offspring, the behaviour of the mother is critical, not only for life itself but for its own adjustment to the environment into which it is born, because the mother is the chief agent responsible for its learning. The mother, by her response to the separate acts of the young, may be said to shape its behaviour.

In this chapter the behaviour of the female before, during, and after parturition is discussed and special attention is given to the development and maintenance of the bond between mother and young. Care of the young, suckling behaviour and development of social relationships by the young up to the time of weaning are also considered.

Unfortunately, much of the available information is restricted to sheep. As far back as 1926 Fraser noted that the most striking feature of a lambing sheep's behaviour was its individuality. "No two sheep behave exactly alike and anything but general prediction is possible." This feature will be discussed later, its importance being that differences in behaviour attributable to breed or to management are very difficult to determine.

PREPARTURIENT BEHAVIOUR

The imminence of parturition in the ewe may be indicated by changes in behaviour such as increased restlessness, pawing and sniffing at the ground, walking around, repeated lying down and standing up (Fraser, 1926; Fraser, 1937; Wallace, 1949; Hersher et al., 1963; Smith, 1965; Arnold and Morgan, 1975). The time prior to parturition at which these changes occur is usually quite short; some ewes may not give any indication at all. Wallace (1949) ob-

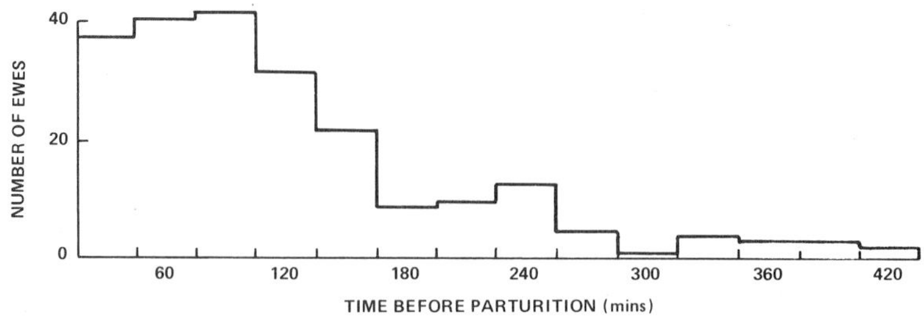

Fig. 5.1. Distribution of time before lambing that behavioural signs of parturition occur. (Source: Arnold and Morgan, 1975.)

served that 15% of ewes in a flock of Romney Marshes displayed uneasiness more than 3 h before parturition, 15% between 2 and 3 h, 30% between 1 and 2 h and 40% within 1 h of birth.

Arnold and Morgan (1975) studied ewes of several breeds, although a majority of them were Merinos, and found that most ewes showed behavioural signs of parturition within 2 h before parturition (Fig. 5.1); in a few cases these occurred up to 11 h before lambing. The distribution of occurrence of behavioural signs with time before lambing was similar to that found by Wallace. There was no change in total time spent walking on the day of parturition in these studies (Morgan and Arnold, 1972). Some 64% of ewes displayed a variety of behavioural signs (Table 5.1), the most common of which were walking in circles, pawing the ground, and getting up and down during labour. Some differences were found between ewes of different breeds; Merino and Southdown ewes pawed the ground less often than ewes of other breeds.

In Suffolk ewes, the onset of parturition is heralded by physical signs such

TABLE 5.1

The percentage of ewes displaying behavioural traits before parturition (Source: Arnold and Morgan, 1975)

Behaviour	Percentage
Stand up during labour	87
Walk in circles before labour	72
Walk in circles during labour	70
Paw at ground before labour	42
Paw at ground during labour	44
Lick herself before labour	6
Lick herself during labour	7
Vocalize during labour	53
Distress calls during labour	32

as protrusion of the amnion, or part of the foetus, or release of amniotic fluids in 48% of ewes, by behavioural signs in 32%, while 20% gave no indication of parturition at all (Fraser, 1968). Fraser's figures for behavioural and physical signs of lambing are quite a bit higher than those of Arnold and Morgan and may indicate that breed differences occur. In Soay sheep a marked behavioural change occurs 4 h before parturition (Shillito and Hoyland, 1971).

No consistent behavioural indications that parturition was imminent were found in cattle by George and Barger (1974). There was a restlessness similar to that found in sheep, with pawing of the ground starting a few days or a few hours before parturition. Many mares walk for hours prior to the onset of labour. As birth becomes imminent they become restless, walking and trotting, pawing, straining, elevating the head and top lip, getting up and lying down (Barty, 1974).

Isolation seeking

Two phenomena widely reported in lambing ewes are isolation seeking and shelter seeking, although there is a lack of quantitative evidence to show the degree to which they occur. Alexander and Arnold (cited by McBride et al., 1967), state that sheep on the verge of lambing have no uniform pattern of behaviour such as isolation or shelter seeking which would be advantageous to the lamb.

Preparturient isolation in domestic sheep has been reported in various breeds in different parts of the world (Fraser, 1926; Fraser, 1937; Hersher et al., 1963; Smith, 1965; McBride et al., 1967; Fraser, 1968). Fraser (1937) states that the first instinct of a ewe about to lamb is to isolate herself. However, McBride et al. (1967) report that, depending on the time of day, the ewe may lamb on the camping site or be left behind by the grazing flock. Only 12% of a Suffolk flock isolated themselves prior to lambing (Fraser, 1968). Alexander and Williams (cited by McBride et al., 1967) observed that Merinos but not Border Leicesters isolated themselves. Thus it appears that there is considerable variation between breeds in the degree to which a lambing ewe will seek isolation. This may be due to a number of factors, including differences in breed and husbandry techniques. Much of the variation reported may be due to differences in paddock size, stocking rate and definition of 'isolation seeking'. It is difficult to assess objectively whether an animal is deliberately seeking to isolate itself from a flock or herd; in many cases it appears that the ewe becomes isolated because she stays at the site where she has spilt amniotic fluid and the flock moves away as it grazes.

Arnold and Morgan (1975) found that 66% of ewes lambed in isolation from the flock and 34% lambed in the area where the flock was grazing or camping. Of the 66%, the majority had in fact sought isolation whilst the rest had been left behind by the flock. The majority of ewes (73%) lambed on the site where the birth fluids spilt. In fact most ewes (63%) did not move away from the site where they commenced labour, and a proportion of those that did move away came back to the site to lamb.

Mountain sheep of America isolate themselves and lamb in inaccessible cliff sites, remaining there until a week or more after lambing (Davis, 1938; Spencer, 1943; McCann, 1956). The risk of predator attack is normally much higher with these sheep than with domestic sheep. It appears that isolation seeking is a behavioural pattern in sheep which, in the process of domestication, has persisted but with some modifications.

In cattle, George and Barger (1974) found that most cows licked the ground where the first discharge of foetal fluids fell and remained at or about that spot until calving was completed. Thoroughbred mares will leave the mob and frequent a certain area, generally only 3—4 days before foaling and, if undisturbed, will foal in this area (Barty, 1974).

Shelter seeking

Shelter seeking, which may be associated with isolation seeking, has been observed by some authors but not by others. Blackface ewes were reported to be more sensitive to climatic conditions near parturition than a month earlier (Hunter, 1954). Welsh Mountain ewes were found to seek shelter at lambing when wind speed was greater than 11 km per hour or the Kata cooling power exceeded 20 millical. cm^{-2} sec^{-1} (Winfield et al., 1969). At higher wind speeds and cooling powers they chose progressively more sheltered lambing sites.

On the other hand McBride et al. (1967) state that near parturient ewes do not display uniform shelter seeking behaviour. In fact they state (citing unpublished data by Obst) that while the ewe reacts to climatic factors such as wind and rain this may lead to lambing in exposed areas, particularly in areas bounded by fences. More recent work in Australia has shown that, when subject to cold stress through shearing, parturient ewes will seek shelter (Lynch and Alexander, 1976).

Prelambing shearing is practised by many farmers in the belief that it will encourage shelter seeking at lambing (Coop and Drake, 1949) but, as McBride et al. (1967) pointed out, there is no evidence to suggest the ewes are seeking shelter for the lambs; it seems that when ewes seek shelter they do so to keep warm.

There are obvious advantages to the new born lamb in the ewe seeking shelter, just as there are advantages to be gained from isolation; it may reduce confusion, swapping and stealing of lambs and provide a better opportunity for the ewe—lamb bond to form. The new born lamb orientates itself towards the nearest object, preferably a moving one (Blauvelt et al., 1960; Smith, 1965). Thus if the ewe is isolated she is likely to be the only object to which the lamb will orientate. Fraser (1968) has said that it helps to prevent lamb stealing. The value of isolation in relation to predation is not clear. It is likely that one of the prime concerns of the mountain ewe in isolating herself in an inaccessible spot is to avoid predators. However, under domestic conditions isolation in open country may make the ewe more susceptible to predatory attack than if she is with the flock.

Preparturient maternal interest

Interest in nearby lambs may occur during the preparturition period (Fraser, 1926; Fraser, 1937; Alexander, 1960; Watson et al., 1968; Shelley, 1970). The reported incidences vary from 0 to 2.2% in mature ewes to 10 and 15% in maidens. This prelambing maternal interest has been observed as much as 2 weeks before lambing, but usually occurs in the 12 h preceding birth (Alexander, 1960). The behaviour may lead to confusion over the ownership of some lambs and rejection by the ewe of either the alien lamb or her own, or of both. Losses due to the interference of other ewes vary: Shelley (1970) attributed only 3% of deaths to this cause whereas in another study 10% of single deaths and 30% of twin deaths were associated with this form of behaviour (Watson et al., 1968).

Twenty-one per cent of all ewes displayed interest in lambs of other ewes prior to the birth of their own lamb in the study by Arnold and Morgan (1975). The incidence of the trait was similar whether the ewe lambed in a large day paddock or a small night paddock and whether the ewe lambed with the flock or isolated from it. Prelambing maternal interest was observed only within 8 h of lambing and on most occasions within 2 h (Fig. 5.2). It varied from a brief inspection to cleaning, suckling and on some occasions to attempted adoption. Most ewes lost interest in the lambs of other ewes immediately prior to the birth of their own lamb.

Four out of 80 ewes persisted in their attention to an alien lamb, and deserted their own lambs (three singles and one set of twins), all of which died. Two of the four ewes adopted and reared the alien lambs whilst the

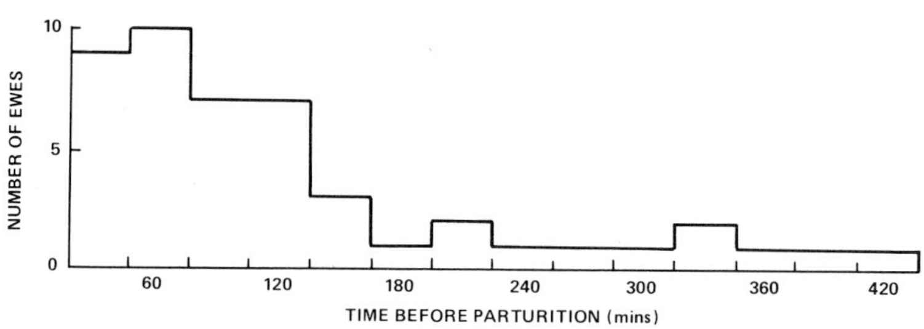

Fig. 5.2. Distribution of time before parturition that ewes show prelambing maternal interest. (Source: Arnold and Morgan, 1975.)

other two deserted the alien lambs which then died. In one other case, persistent prelambing interest led to two ewes exchanging their lambs and each successfully rearing the alien. Thus four lambs were reared by mothers that were not their own.

Little is known about calf stealing among cattle, but foal stealing is likely to occur if a dominant mare has a dead foal. She will steal the foal of a subordinate mare (Barty, 1974).

The birth process

The birth process has been described by Collias (1956) and Hersher et al. (1963). The ewe usually lies down during labour, but may stand during the final stages of expulsion (Plate 5.1). Labour is generally short, most lambs being born within an hour of first protruding at the vulva (Wallace, 1949; Alexander, 1960). However, as Fig. 5.3 shows, quite a number of ewes have prolonged labour due, usually, to dystocia. Fraser (1968) showed a similar curve for Suffolk ewes. Labour is much shorter for second born lambs (Arnold and Morgan, 1975). The interval between births of the first and second lamb varied (Fig. 5.4) but on average was $<$45 min.

Young ewes may have slightly longer labour than old ewes (Alexander, 1960); however, the difference does not appear to be of major importance.

Plate 5.1. Ewe cleaning one lamb whilst its twin sucks.

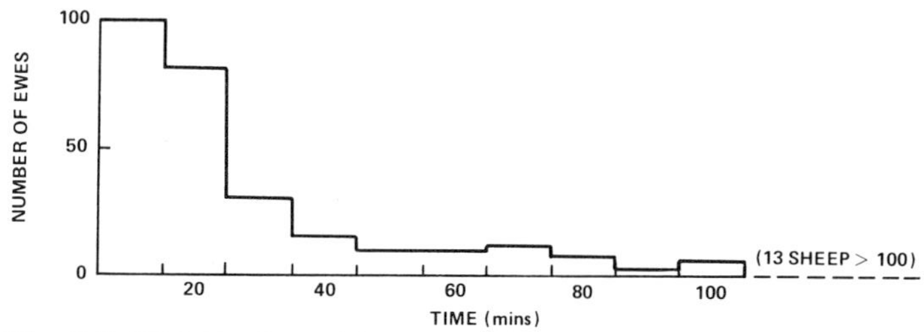

Fig. 5.3. Distribution of length of labour in ewes. (Source: Arnold and Morgan, 1975.)

Alexander (1960) also concluded that a small difference between age of ewes in length of labour was due to differences in nutritional levels and birth weights of the lambs. A comparison of Alexander's (1960) and Wallace's (1949) results suggests that there are only small differences due to breed. Because of the large variation which occurs in the length of labour, differences of less than 30 min are probably of minor biological significance only.

Difficult births appear to make up only a small proportion of all births (Gunn, 1968), but mortality is high among lambs subject to difficult births; 89% of lambs in unassisted difficult births died in Gunn's study. He observed that the incidence of difficult births is affected by breed, type of birth, birth weight and sex of lamb, and it also varied between years. He suggested that the breed differences which he observed (higher incidence among Cheviots than Blackfaces) may be partly an expression of fundamentally different behaviour at parturition, and therefore a resulting variation in the amount of attention by the shepherd.

The distribution of births over the 24 h seems to vary with breed of sheep

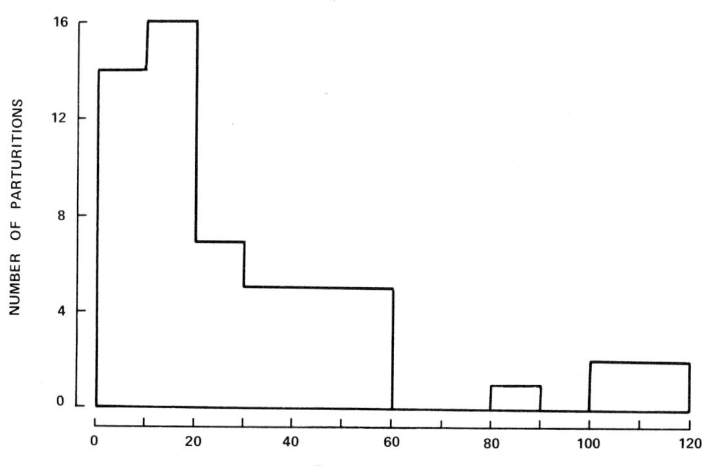

Fig. 5.4. Distribution of intervals between births of twin lambs. (Source: Arnold and Morgan, 1975.)

as is shown in Fig. 5.5. The comparisons given are for sheep lambing in the paddock and when daylength was similar in each study. Suffolk ewes had a major peak of lambing between 1600 and 2000 h and none lambed between 0900 and 1200 h. By contrast, the Merinos studied by Arnold and Morgan had only one small peak between 1700 and 2100 h. Wallace confined his

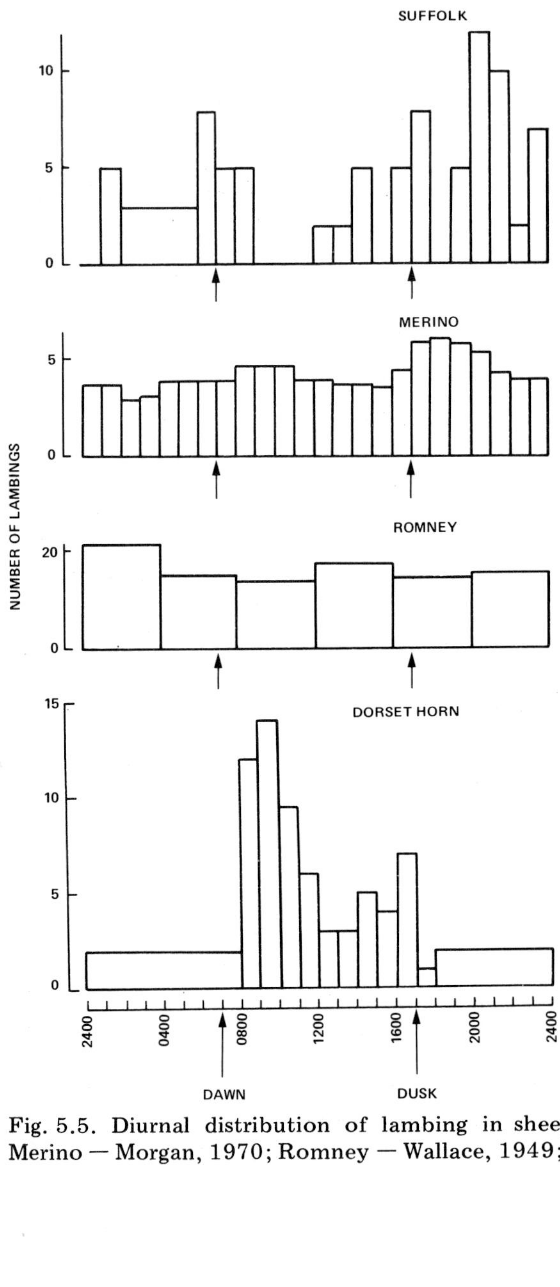

Fig. 5.5. Diurnal distribution of lambing in sheep. (Sources: Suffolk — Fraser, 1968; Merino — Morgan, 1970; Romney — Wallace, 1949; Dorset Horn — George, 1969.)

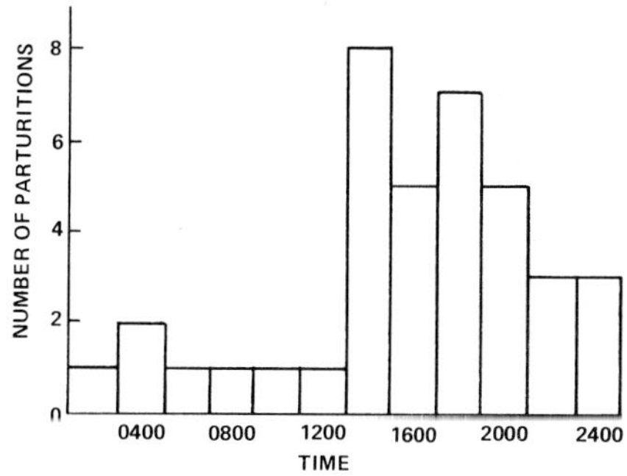

Fig. 5.6. Diurnal distribution of calving in cows. (Source: George and Barger, 1974.)

Romney ewes to pens at night and used lighting so the results may have been influenced by this.

The most detailed comparison between breeds available is between Merinos and Dorset Horns lambing at three different times of the year (George, 1969). Although observations were made only during daylight, the Dorset Horn ewes had a marked peak of lambing between 0800 and 1200 h, whereas Merinos had no discernable peaks. This and the patterns cited would seem to be genetically determined since daylength was not a factor.

Selman et al. (1970a) found that 53% of calves were born to their hips whilst the dams were recumbent. At this stage the dams rose and dropped their calves shortly afterwards. The remainder calved whilst lying.

In housed cows, Arthur (1961) recorded that 66% calved between 1800 and 0600 h, while Ewbank (1963) obtained, with a small group of Ayrshire cows, a figure of 56% for this period. However, Hereford cows calving in a paddock calved more often during the period 1400—2400 h than at other times of the day (Fig. 5.6) (George and Barger, 1974).

In mares, the birth process is rapid (5—47 min) and 98% of mares give birth in the lying position (Rossdale, 1968). They usually remain in that position for 30 min after giving birth.

DEVELOPMENT OF MATERNAL CARE AND NURSING

Cleaning and attraction to the newborn

Although maiden ewes may be slightly slower than old ewes, the ewe usually commences cleaning her lamb within a few minutes of birth (Collias, 1956; Alexander, 1960; Hersher et al., 1963; Arnold and Morgan, 1975). Most cattle behave similarly (Selman et al., 1970a), with continuous licking proceeding for up to 1.5 h; adult cows were found to average 40 min but heifers only 11 min. The mare will lick its foal for several hours.

A delay in the newborn lamb getting to its feet or in the onset of grooming is associated with a long labour (Arnold and Morgan, 1975). Most ewes and cows commence grooming at the head (Table 5.2) and work down the body. However, if the lamb or calf starts to stand or to call, the dam will often return to licking the head.

The birth membranes are normally eaten during the cleaning process: the placenta is usually voided more than three hours after lambing and only a small portion of it is eaten by about a third of ewes (Arnold and Morgan, 1975). George and Barger (1974) found that 14 of 17 cows (82%) ate the placenta, but mares rarely do so (Rossdale, 1968).

During grooming both dam and offspring vocalise and this is important in the development of the parent—offspring bond. Selman et al. (1970a) report three distinct sounds made by cows. During initial licking loud bellows were made. Later soft pharyngeal grunts were made with the mouth closed. A similar sound to this, but louder was heard when a calf wandered away from its dam. The ewe has a low pitched gurgling call and the doe a series of low pitched and short bleats (Collias, 1956) which they use when nursing their offspring. The distress call in both species is a relatively prolonged, loud and high pitched 'scream'. Shillito and Hoyland (1971) have measured these calls acoustically in Soay sheep and found that individual sheep have characteristic voices.

Experiments with sheep and goats suggest that the dam is initially attracted to her young by the smell and taste of the birth fluids (Fraser, 1926; Collias, 1956; Klopfer et al., 1964). It has been suggested that the ewe's appetite (Hersher et al., 1963) and thirst (Collias, 1956) may also be involved. Alexander (cited by McBride et al., 1967) prevented ewes from smelling the birth fluids during parturition and then presented them with various objects including a newborn lamb. The reactions of the ewes were variable, but interest was displayed only in objects which were wet and the greatest interest was shown in the new born lamb. Collias (1956) was able to attract a ewe to follow her new born lamb (which was quite motionless) although she had not had prior contact with it. It was also possible to attract

TABLE 5.2

Sequence of grooming (percentage occurrence) (Source: Arnold and Morgan, 1975)

Sequence of grooming (n = 357)							
First		Second		Third		Fourth	
Head	65	Head	12	Head	12	Head	11
Fore quarter	9	Fore quarter	58	Fore quarter	24	Fore quarter	9
Body	13	Body	23	Body	56	Body	7
Hind quarter	13	Hind quarter	7	Hind quarter	8	Hind quarter	72

the ewe for short distances with a rag which had been rubbed in the birth fluids, but not with a rag soaked in water. Washing the birth fluids off with detergent did not prevent the ewe from accepting her lamb (Collias, 1956). He concluded that the birth membranes facilitate, but are not essential to the initiation of maternal care. Thus it appears that the initiation of cleaning is probably an innate form of behaviour which may be augmented by olfactory and gustatory stimuli from the birth fluids.

The establishment of the ewe—lamb bond in multiple births follows the same pattern except that the grooming of the first lamb may be interrupted by the birth of the second. However, the ewe, if distracted by the pain of parturition, may wander off to give birth to the second lamb and abandon the first (McBride et al., 1967).

Initiation of suckling

Like the new born of most ungulates lambs get to their feet soon after birth, the time taken varying from <10 to >180 min (Arnold and Morgan, 1975), 70% standing within 30 min (Wallace, 1949; Alexander, 1958). Cleaning by the ewe appears to stimulate the lamb to stand (Alexander, 1960; Alexander and Williams, 1964).

There appears to be little or no relationship between breed and time taken to stand and drink, length of labour, type of presentation, age of ewe, birth weight, plane of nutrition and type of birth (Wallace, 1949; Alexander, 1958; Arnold and Morgan, 1975). As with length of labour, the time from birth to stand and drink (Fig. 5.7) varied greatly, and small differences in mean values for breeds are of doubtful significance.

Lambs are also quick to suck, the interval from birth to first drink varies from 20 min to more than 3 h, but 60—80% suck within an hour of birth (Wallace, 1949; Alexander, 1958; Arnold and Morgan, 1975). The Romneys observed by Wallace (1949) and Merinos observed by Alexander (1958) sucked more quickly than the Corriedales observed by Alexander (1958). There were only minor differences due to plane of nutrition and type of birth in these studies. Arnold and Morgan (1975) found no consistent effects of weather conditions at the time of lambing and the time taken by lambs to get their first drink in three quite different environments in Australia, nor was there a consistent difference in the time taken by single and twin born lambs. Within a given lambing period they found that air temperature, wind speed and rainfall during the period had no effect on the interval from birth to the lamb obtaining its first drink. The lack of a statistically significant relationship reflects the enormous variation between individuals. Under experimental conditions, Alexander and Williams (1966) found that teat seeking activity was depressed under cold, wet conditions, and ceased when rectal temperatures fell below 37° C.

Calves usually remain passive for the first 30 min after birth (George and

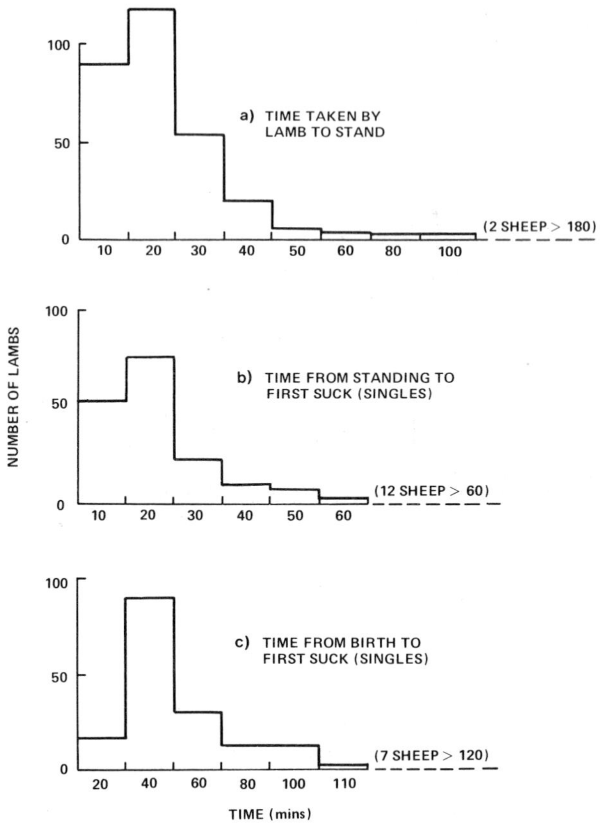

Fig. 5.7. Distributions of times taken by lambs to stand after birth, and to drink after standing. (Source: Arnold and Morgan, 1975.)

Barger, 1974), and these workers found an average interval of 111 min from birth to first drink. Considerable differences have been found between calves from beef and dairy cows (Selman et al., 1970b). Calves of beef cows stood quicker than calves from dairy cows (35 vs. 58 min) and had their first drink quicker (81 vs. 261 min). Rossdale (1968) found that foals took 30 min, on average, to obtain their first drink.

The teat seeking drive in the newborn animal is strong and it will try to suckle any part of the dam's body. Thus the route to the udder is variable. Poindron (1974) found that most lambs start from the head of the ewe, and stop at the shoulders and try to suckle at the axilla first. Teat seeking in calves is directed to the fore or hind quarters, with the calf trying to thrust its head as high as possible into a recess (Selman et al., 1970b). Thus calves with dams that had large pendulous udders took much longer to their first suckling (40 min) than calves with dams that had small udders (17 min).

Since beef cows had smaller udders, the differences in time taken to obtain the first drink was due, in large part, to the udder shape.

The movement of the ewe, her cleaning of the lamb and tendency to turn her body as she continues cleaning may help to orientate the lamb towards the udder (Hersher et al., 1963). The ewe may also lift a hind leg to assist the lamb in its search. Grooming and directional orientation are not essential for the lamb to find the udder, but they facilitate the process (Alexander and Williams, 1964). Neither alone has a significant effect.

Initially the cow will face the calf, continuing to lick it, and it takes about 20 min before it stands still for the calf's teat seeking advances (Selman et al., 1970a). Once a drink has been obtained, subsequent teat seeking time is much shorter e.g. 5 min for the second compared with 17 min for the first suckling in calves (Selman et al., 1970b). The cow may help the calf seeking the udder by moving forward as the calf moves along the cow's flank. The mare helps the foal in its approach to the udder by resting the hindleg and contracting the abdominal muscles.

BOND FORMATION AND MATERNAL CARE

The critical period

The 'critical period' after parturition during which a mother will form a bond with her offspring is short. In sheep and goats this period seems to be less than 5 h (Collias, 1956; Smith et al., 1966) and in cattle Le Neindre and Garel (1976) suggest that it is less than 3 h from parturition. After this, if there has been no contact since birth with any young, there is a rapid decline in maternal interest in the new born with increasing aggressive acts against the young when it approaches the dam. However, only a short period of contact (5—30 min) is needed to establish a bond which enables the mother to discriminate its own offspring from alien offspring (Klopfer et al., 1964; Smith et al., 1966; Klopfer and Klopfer, 1968). It appears that in this initial period a ewe, doe or cow that has given birth approaches any young animal and will become attached to it. This explains why lamb swapping occurs when two ewes lamb close together at the same time (Welch and Kilgour, 1972; Arnold and Morgan, 1975). The attachment is learnt and is not imprinted as once thought (Klopfer and Gamble, 1966).

After this critical period the young will be accepted after several hours of separation (Collias, 1956). Also, it is during this receptive period that alien young can be readily fostered onto a female goat (Klopfer and Klopfer, 1968). Fostering after this period is possible but usually requires enforced contact between dam and young and the prevention of butting by the dam (Hersher et al., 1963).

Recognition of the offspring by the dam

Nearly all the work on this aspect has been done with sheep but probably applies with a few qualifications to other species. There are two aspects of importance; one is the identification of the offspring when the dam is in close contact with one or more young animals, and the other is the recognition of offspring from a distance of 10 m or so, whereby the mother goes directly to its offspring.

The ability to identify its own offspring from aliens is developed within hours of parturition and is dependent on the odour of the lamb (Tschanz, 1962; Bouissou, 1968; Morgan et al., 1975). The identification is usually made by smelling the anal-genital area.

After a week more than 50% of ewes can recognise their lambs from a distance of 12 m (Morgan et al., 1975). It seems unlikely that the odour of the lamb could be identified at this distance. Both visual and auditory cues are important because without them only 9% of ewes recognised their lambs at this distance (Morgan et al., 1975). These workers concluded that the relative importance of the senses was sight > hearing > smell and this has been substantiated by other work (Alexander and Shillito, 1977a).

Tschanz (1962) found that auditory cues did not have specific value in identification of lambs. Merino ewes show a significant preference for the voices of their own lambs and most can locate their lambs when they were in a yard 25 m away (Poindron and Carrick, 1976), but this could also involve visual cues. The visual cues for the ewe come mainly from the lamb's head (Alexander and Shillito, 1977a).

Barty (1974) notes that the mare's ability to identify her own young immediately after birth is very poor. A 3—4 day old foal, dry and having an alien odour, can be substituted immediately for a foal born dead and is generally readily accepted by the mare.

Recognition of the mother by the offspring

The lamb's ability to recognise its mother develops rapidly over the first week of life (Fig. 5.8). As it learns then its approaches to other ewes decrease and Shillito and Alexander (1975) found that only 2% of lambs of five breeds attempted to suck alien ewes 24 h after birth. Calves and foals are quite different and will attempt to suck alien cows or mares if the opportunity arises.

Visual cues are important to the lamb in recognising its mother from a distance, as shown by Shillito (1975). She found that lambs of Clun Forest, Finnish, Jacob, Dalesbred and Soay breeds of sheep could identify their mothers from two other ewes at 16 m and run to them: if they could not see their mothers about 30% failed to find them.

In Merino lambs, the importance of visual cues decreases in the second

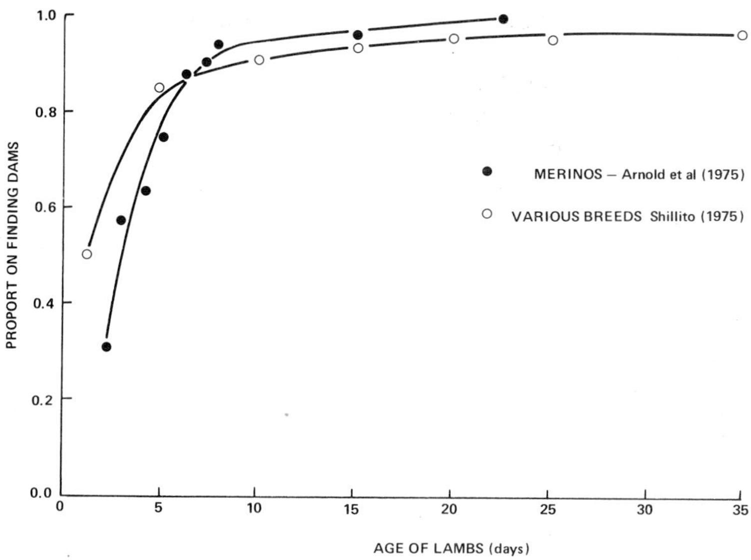

Fig. 5.8. Change with age in the ability of lambs to find their dams from a distance of 10 m.

week of life and auditory cues became much more important (Fig. 5.9) (Arnold et al., 1975). In that study when the lambs could neither see nor hear their dams, only 10% found them even though they were only 6 m away. When deprived of auditory cues most lambs still sought their mothers but many failed to identify them. This contrasts with sheep of the breeds

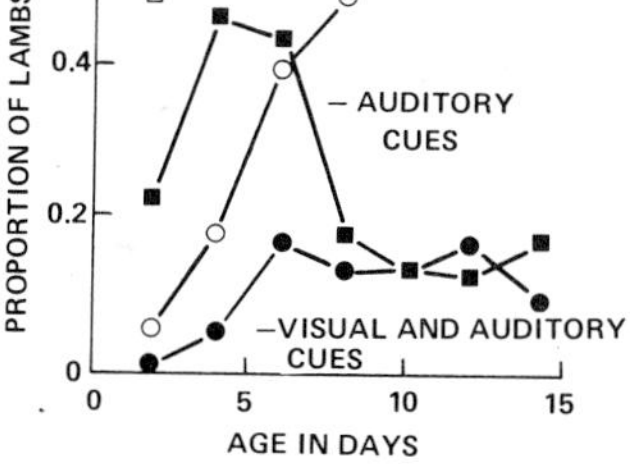

Fig. 5.9. Effects of sensory impairment on the ability of lambs to recognise their dams. (Source: Arnold et al., 1975.)

studied by Shillito who found that many lambs did not search if the ewes were silent.

Mutual recognition

To study the role of the senses in recognition, most workers have restrained either the mother or the offspring and measured the response of the other. In reality, the two may both be seeking each other if separated. A study of mutual recognition in Merinos over the first month of life (Alexander, 1977) showed that both visual and auditory cues were important when ewe and lamb were 13 m apart. Auditory cues were of secondary importance if strong visual cues indicated an alien partner.

From observations with sheep it is clear that in the first week of life the lamb tends to stand still and call and the ewe does the active searching. Progressively, with age, the lamb does more and more of the searching and the ewe stands still. This pattern remains because the matrilinearly related social groupings of animals depend on the offspring retaining its attachment to its mother and not vice versa.

Under rangeland conditions in Oklahoma, Dwyer (1961) observed that calves did not seek their dams but rather the cows sought the calves. Cows and calves seemed to identify each other by smell and voice more than by sight. The cows called for their calves and the calves answered from a considerable distance, then they walked toward each other. Wagnon (1963) reported similar observations, but neither author says whether this pattern holds at all ages of the calf.

Social attachment between mother and young

The reason for differences between sheep and cattle is that in sheep the mother—infant relationship occurring after parturition is of two types referred to as 'hiders' and 'followers' (Lent, 1974).

With the 'hiders', which include cattle and goats, mother and infant offspring are separated and out of contact for long periods of time. Infants of hider species tend to act independently in the selection of hiding sites, but are dependent, at least initially, upon their mothers for initiation of activity periods. They are dependent, also, on their mothers finding where they are hidden. By contrast, sheep are 'followers' and the young must keep with the mother for maintenance of the bond.

There is considerable difference of opinion as to whether calves are left in 'nurseries' with guard cows. All the reports of nurseries with guard cows come from observations under extensive rangeland conditions. Even so, opinion varies on whether there is a deliberate organisation of the guarding.

During the first few days of life the cow will spend much of her time close to the calf but will go for water, leaving it behind. Within the first week of life the calf will begin to follow the cow for periods each day but groups of calves will be found lying together for much of the time while the cows are grazing. It is in this period, before the calves are themselves grazing for any length of time, that nurseries are said, by some, to be formed.

The authors have observed that in paddocks of up to 50 ha the groups of calves are usually unaccompanied; only occasionally are one or two cows with them (Plate 5.2).

Observations in Queensland by Rankine and Donaldson (1968) are that a cow does not really seek to protect her calf by hiding it; the calf just flops down anywhere, often as if for concealment in long grass, but quite often out in the open. From about 6 weeks calves begin to congregate in groups of 20—25, usually with only one or two cows near them. "Sometimes, a single cow near them gives rise to the story of 'nursemaid' but there does not seem to be purposeful behaviour of this sort." On the Barkly Tablelands in arid Northern Australia, nursery areas were observed 1—2 km from water with up to 10 calves left with one or two cows (Schmidt, 1969). By contrast with Rankine and Donaldson, Schmidt says that calves of more than 1—2 weeks of age were rarely left in nursery areas. However, in neither study were individual calves identified so that it is difficult to assess whether this difference is of real significance.

In California, Wagnon (1963) observed that as the calf became more active its dam usually joined with other cows that had calved. When the cows left

Plate 5.2. A nurse cow?

their calves to graze they "usually rotated their absence so that there would be at least one cow with the calves at all times". The exception was when supplements were being fed. The 'nursemaid' cow usually came running to any calf that bawled.

The problem is that there have been no objective studies of the question. There would be obvious survival value in having calves 'guarded' especially when predators are a problem — as occurs in Northern Australia with the dingo and in North America with various predators.

The young lamb is more precocious than the young calf, following its dam on the day of birth, often for quite long distances. This can be fatal in hot climates (Morgan et al., 1972b). The ability of Merino lambs to follow their mothers was studied in two locations (Arnold, unpublished). Each day the flock was disturbed by a man with a dog which ran about disturbing the sheep. As seen in Fig. 5.10, 40% of lambs in one flock ran off with their dams when a day old, and 56% in another. The reactions of the ewes differed strikingly in the two flocks, with few of those in flock B being prepared to stay with their lambs when disturbed even if the lambs could not keep up with them. Thus an average of 15% of ewes in this flock simply deserted their lambs when disturbed, compared with 5% in flock A. The difference between the flocks was in the amount of experience of man they had. Flock A was handled at least every month whilst flock B was handled only twice a year.

The response of the ewe to a disturbance reflects only in part the lamb's ability to stay with the dam if she moves. There is an association between willingness to stay put when disturbed, and the birthweight of the lamb (Fig. 5.11). To us this indicates a stronger maternal instinct in ewes that are well

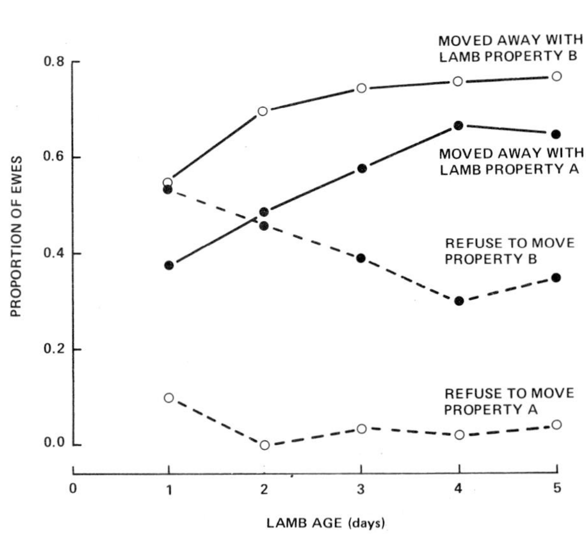

Fig. 5.10. Changes with age in behaviour of ewes on two properties to disturbance.

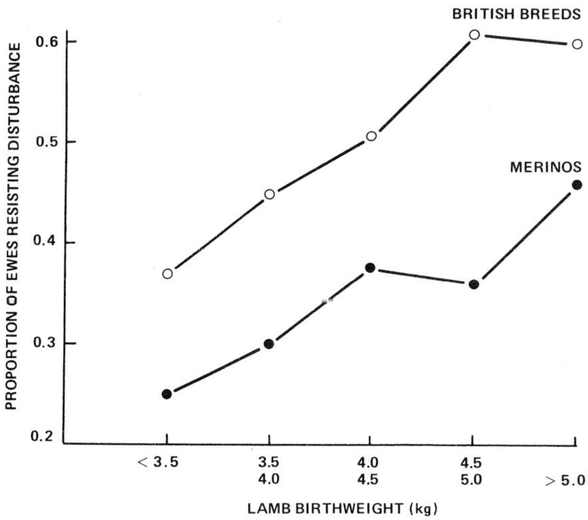

Fig. 5.11. Differences between breeds of sheep in resistance to disturbance in the week after lambing in relation to the birth weight of lambs.

fed and produce big lambs. There are distinct breed differences in this aspect of maternal behaviour, with more ewes of British breeds of sheep than Merino ewes resisting disturbance. This has also been found in New Zealand (Whateley et al., 1974). Age also has an effect, ewes lambing for the first

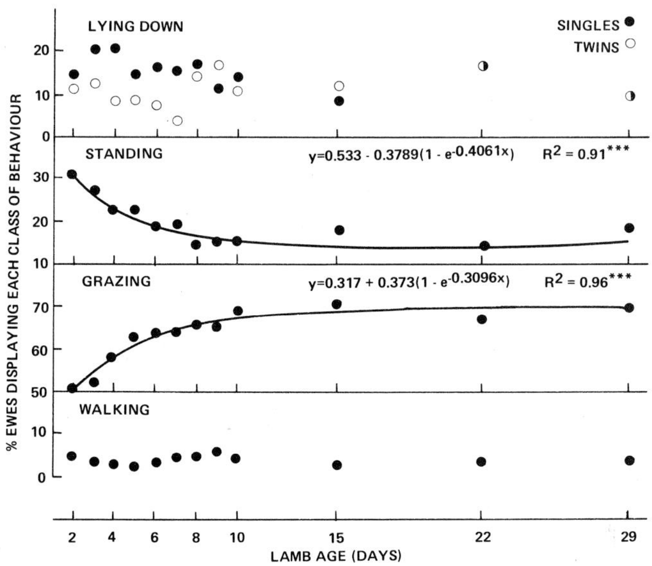

Fig. 5.12. Changes in the activities of ewes with time after lambing. (Source: Morgan and Arnold, 1974.)

time being more easily disturbed than those that have lambed before.

It is interesting that ewes with twins resist disturbances more often than those with single lambs. This indicates either that twins, being smaller, are less able to follow their dams, or that the ewe with twins is a better mother.

The behaviour of the ewe and lamb are, of course, related. It takes 10 days after lambing for the ewe to get back to a normal amount of grazing time (Fig. 5.12). However, the time spent by the lamb in different types of behaviour changes continuously over the first month of life (Fig. 5.13). For example, it spends increasingly less time lying down and more time grazing. Play also increases, being at a maximum at day 10.

The behaviour of the lamb, however, is not independent of that of the ewe. Playing mainly occurs when the dam is more than 25 m away grazing (Morgan and Arnold, 1974).

Maternal care seems to involve not only suckling the lamb and staying close to it to ensure that the lamb does not become lost, but also ensuring

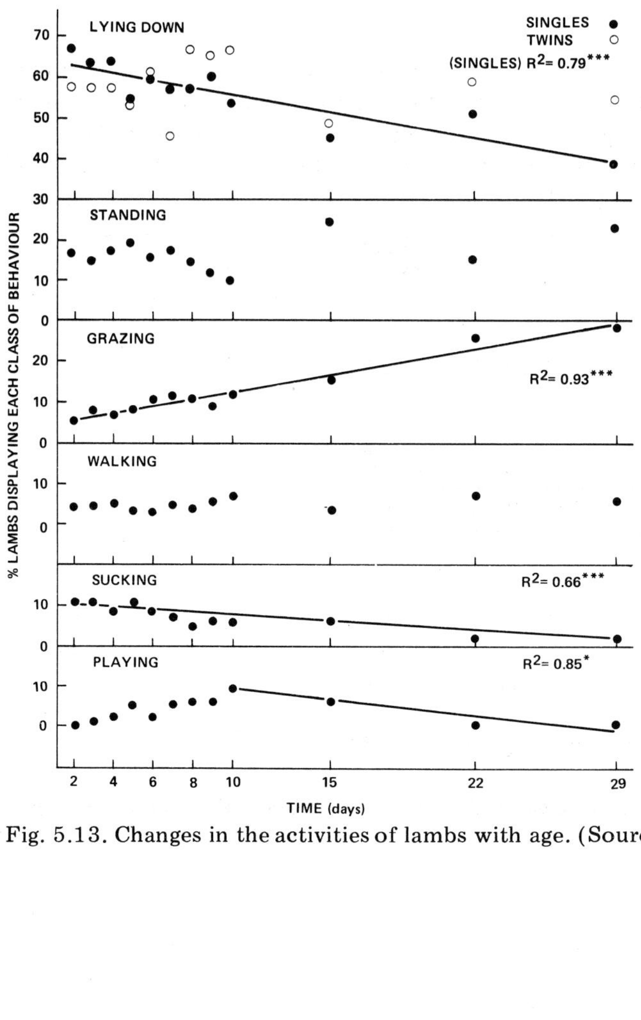

Fig. 5.13. Changes in the activities of lambs with age. (Source: Morgan and Arnold, 1974.)

that if the flock moves the lamb maintains its contact with its dam. This suggestion is derived from several facts obtained by Morgan and Arnold (1974). When the ewe was lying the lamb was lying on 93% of occasions and this did not change with age, nor did the distance between them, which was usually less than a metre. Even when peer groups were formed, lambs were lying at the same time that their dams were lying. This relationship between ewe and lamb is not a flock reaction since there were never more than 22% of ewes lying at any one time. When the ewe was standing the lambs stayed close by, either standing or lying. Also, if the ewe was walking, the lamb was usually walking. Independent activities like playing occurred mostly when the ewe was grazing.

The attachment between ewe and lamb is maintained as the lamb gets older, despite the development of independent activities; it is manifest in the close relationship of lamb behaviour to ewe behaviour and spatial proximity when the latter lies, stands or walks. The behaviour patterns observed provide the necessary basis for maintenance of social organisation within the flock as discussed in Chapter 2.

Suckling behaviour

In early lactation lambs are allowed to suckle at any time (Munro, 1956; Ewbank, 1967b) and for as long as they wish (Munro, 1956). However, after 1—2 weeks the ewe starts to restrict the frequency and duration of suckling periods (Fig. 5.14), generally by walking away from the lamb as it attempts to suckle (Munro, 1956; Ewbank, 1967b). Ewes with twins have been seen to reject suckling attempts more often than those with singles (Ewbank, 1967b). The behaviour of the lamb following an unsuccessful attempt to suck varies: Munro (1956) observed that an unsuccessful attempt was followed by a successful one a few minutes later, whereas Ewbank (1967b)

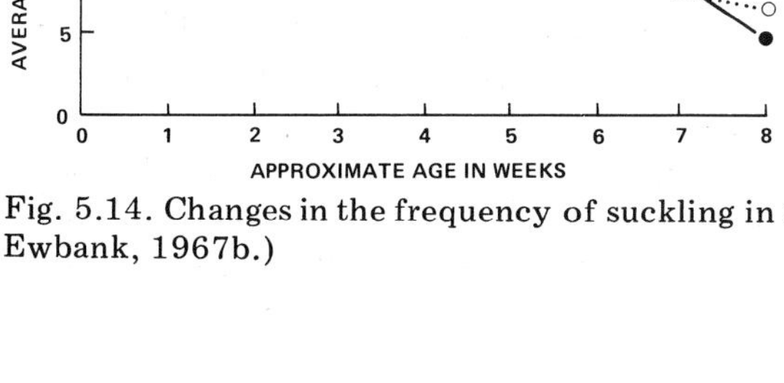

Fig. 5.14. Changes in the frequency of suckling in single and twin lambs with age. (Source: Ewbank, 1967b.)

Plate 5.3. Lamb suckling.

found that it was usually an hour before a lamb attempted to suckle again.

Information on the diurnal variation of suckling is scant, although suckling appears to be evenly distributed between the day and night (Sojetado, 1952). However, the amount of milk drunk may vary throughout the day in the tropics (Moule and Young, 1961), intake being least in the hottest parts of the day, which it was suggested was due to reduced milk production.

Ewes with twins usually allow one lamb to suck in the absence of the other in the first week of lactation only. It appears that by the eighth week some lambs had learnt this (Ewbank, 1964, 1967b) because on some occasions if one twin arrived before the other it would run back to meet it, both then approaching the ewe together.

As the ewe becomes increasingly reluctant to allow the lamb to suckle there is a decline in the amount of time spent in suckling (Sojetado, 1952; Ricordeau et al., 1960; Brown, 1964; Ewbank, 1964, 1967b). This is made up of a decline in both the frequency and the duration of suckling. The lamb must therefore increase its efficiency of suckling in order to maintain milk intake.

Ricordeau et al. (1960) reported no differences between the suckling frequencies of twins and singles. However, Munro (1956) and Ewbank (1964, 1967b) found that twins suckled much more frequently than singles in early lactation; gradually the difference in frequency narrows until it disappears at the fourth week (Ewbank, 1964, 1967b). The difference in frequency of suckling between singles and twins may be due to allelomimetic behaviour or social facilitation — one twin may wish to suck and so the other copies it, with the reverse occurring at another time (Hafez and Lindsay, 1965). In this way twins would drink more often than singles. On the other hand, differences in milk production may be responsible (Ewbank, 1967b). Twin-bearing ewes produce more milk than those with singles but do not produce twice as much. Thus the twins may be more hungry, leading to more frequent attempts to suckle. Ewbank goes on to suggest that the decline in suckling frequency of the twins to the level of the singles may be due to the ewes not allowing unlimited access to the udder, or the lambs finding alternative sources of feed thereby reducing hunger, or both.

The frequency of suckling reported by different authors varies greatly, but because of differences in the timing of their observations it is not possible to determine whether breed differences occur. There is no difference between singles and twins in the duration of suckling.

Like some other mammals, lambs appear to develop a preference for a particular side of the ewe when suckling (Wallace, 1949; Brown, 1964; Ewbank, 1964, 1967b). However, they are relatively slow in developing the preference, not showing real preference until the fifth week (Ewbank, 1967b). Single lambs which display a preference suck both teats from the same side, thereby preventing uneven distension of the udder and reducing the risk of mastitis. Fewer twins than singles develop side preference (Ewbank, 1964, 1967b). However, the numbers involved in the work were not sufficient to make a conclusive statement. Before developing side preference twins may stop partway through suckling, swap sides and recommence suckling. There are unsubstantiated reports that twin lambs always suck from a particular side (Cowlinshaw, cited by Hafez and Scott, 1962; Brown, cited by Ewbank, 1964).

It has been suggested that following the removal or loss of a twin, the remaining twin may continue to suck from one side only (Brown, 1964). This would result in incomplete utilisation of the milk supply and possibly lead to uneven distension and mastitis. However, it has been found that if one twin is removed, within 4 h the other utilises both sides of the udder (Ewbank and Mason, 1967).

Milk production has been found to differ between sides in some ewes, the difference being greater in ewes with singles than those with twins (Moore, 1966). Thus, if the ewe is fully stimulated by suckling twins, the differences indicate differences in potential for milk production. The greater difference in ewes with singles indicates that the lamb probably develops a preference for the better side.

Fewer studies have been made of suckling behaviour in cattle. Walker (1950) observed that suckling frequency was six to eight times a day at

Plate 5.4. Calf suckling.

3 days of age and that suckling was done at night by young calves only if darkness lasted more than 5 h. However, Wagnon (1963) (Fig. 5.15) found that in older calves suckling occurs at all hours of the day and night, but most often between 0500 and 0600 h, with suckling periods lasting 1—27 min (average 10.4 min), giving a daily total of 49 min from four suckling periods.

In another study, Angus and Hereford calves were compared from about 30 days of age to 170 days of age (Walker, 1962). Suckling frequency did not change, averaging four times a day for the Angus calves and 4.5 times for the crossbred calves. The behaviour of Angus and Friesian × Angus calves from birth to 120 days of age was observed from dawn till dusk by Nicol and Sharafeldin (1975). They may have missed some nocturnal suckling because the frequencies are lower than found by Walker (1962), but the changes in frequency (Fig. 5.16) obtained are probably valid. They found that suckling was initiated by both cows and calves, either by the calf approaching the mother or from a call by the mother. Suckling frequency declined rapidly from 5.6 to around 3.5 times a day by 25 days of age. This change was accompanied by an increase in time spent at each suckling from 6 min to 10—11 min, so that total suckling time remained constant. This was despite a rapid increase in time spent grazing. Friesian × Angus calves averaged 15% more suckling time a day, indicating a heterosis effect in which a greater suckling drive must have been a part.

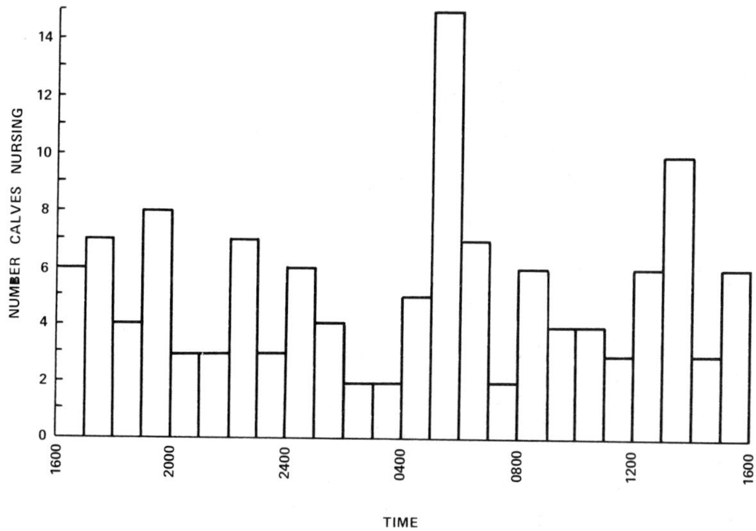

Fig. 5.15. Diurnal distribution of nursing in calves. (Source: Wagnon, 1963.)

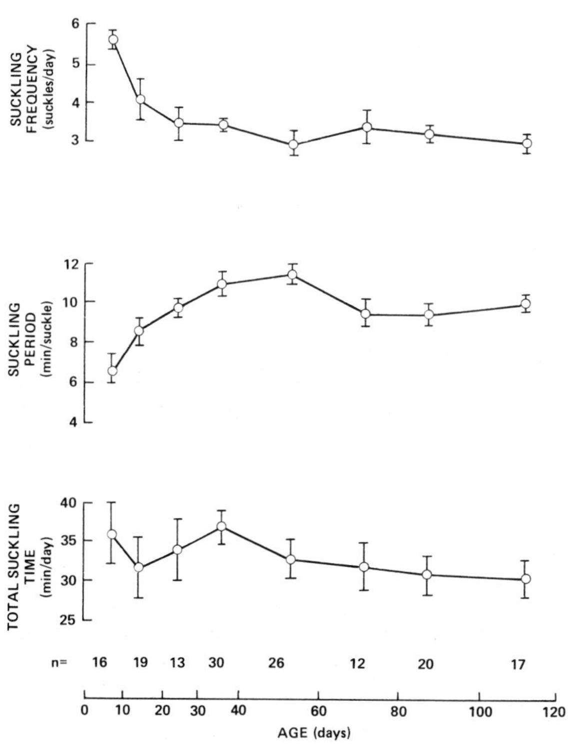

Fig. 5.16. Changes in the frequency of suckling and in suckling time in calves. (Source: Nicol and Sharafeldin, 1975.)

Plate 5.5. Mare with foal.

In multiple suckling systems used in cattle production Kilgour (1972) has found that cows with three or four calves average only 11.5 sucklings a day. However, in a group of seven cows and 24 calves, each cow allowed between 10 and 17 different calves to suckle over a 5 day period.

There is only one reported study on horses. Schoen et al. (1976) observed that foals were nursed mainly between 0800—1000 and 1800—1900 h and that each period of suckling averaged 3 sec. However, these workers only observed the horses during daylight.

Defence of the young

The ewe, like the female of many mammalian species, while normally timid and non-aggressive often becomes aggressive in the defence of her lamb. Hornocker (1969) observed three bighorn ewes chase off a bobcat that was chasing a lamb. Ewes with lambs at foot have been observed to repel foxes and ravens (Alexander et al., 1967). While foxes do not appear to be particularly aggressive (Alexander et al., 1967; Mann, 1968), ravens are very persistent in their attacks (Smith, 1965). It has been suggested that lambs may be more susceptible to predatory attacks if they become separated from their dam. However, ewes and their lambs usually 'mother up' at the first sign of a disturbance. The low level of primary predation in Southern Australia (McHugh and Edwards, 1958; Davies, 1964; Dennis, 1969) suggests that, although the incidence of predatory attack is probably low, the ewe defends her lamb with a fair measure of success.

W.A. Low (personal communication, 1977) has observed cows in Central Australia driving dingos away from their calves, and Rankine and Donaldson (1968) report that when a calf is attacked by dingos "its terrified calls bring any cattle within hearing, even bullocks, galloping up and this frightens off the dingos".

Weaning

In husbanded flocks and herds the husbandry practice of 'weaning' involves the physical separation of dam and offspring. This breaks the social attachment between mother and offspring and will also deprive the young of the mother's milk if the dam has not naturally weaned her offspring.

In sheep the ewe weans the lamb by preventing suckling. The age at which this occurs depends on the ewe's milk supply (Arnold et al., 1979). The larger the milk supply, the later the weaning. In most cases weaning is complete by 120 days, but some ewes will not wean their lambs until 180 days. The process is completed within a week and, once complete, the ewe will not allow the lamb to suckle. After weaning, the lamb, if disturbed, will continue to call its dam, the dam may answer the call and the lamb will attempt to suckle. Gradually this also will cease and, although an attachment remains, vocalisation ceases. At this stage the attachment depends on the young and if there is no physical separation of the weaned lambs from the ewes this will develop, in the case of females, into matrilinearly related social groupings as described in Chapter 2.

ABNORMAL BEHAVIOUR

There are several behavioural traits that can be classed as abnormal. We do not include pre-parturient maternal interest in other young unless it leads to temporary or permanent adoption, which more commonly occurs after parturition. Desertion and various types of poor maternal behaviour have to be considered.

If the ewe takes more than a few minutes to stand after giving birth and to start cleaning her lamb, the lamb may occasionally wander off. The most common traits of poor maternal behaviour are butting of the lamb in the first few hours after birth — this was observed in 5% of primiparous Merino ewes by Alexander (1960) — and failure of the ewe to stand still whilst the lamb is seeking the udder. In the same study Alexander reported that 29% of ewes exhibited this latter trait whilst Arnold and Morgan (1975) found it in 15% of ewes. It usually ceases within a few hours, but the effect is to delay the lamb getting a drink, which under extremes of heat and cold can be fatal.

There is usually a low incidence of interference by ewes that have lambed or are about to lamb, although it was found to be 33% in one flock of Corriedale ewes (Winfield, 1970). The consequences are adoptions or the separation of lambs from their dams either temporarily or permanently. The incidence of adoptions in 400 ewes studied by Arnold and Morgan (1975) was 3%, and Shelley (1970) obtained a similar figure.

Desertion varies from a complete lack of interest being shown in the lamb after birth, with the ewe walking away from the site of lambing (Plate 5.6), to temporary desertion for a few minutes or a few hours. Winfield (1970) reported that 22% of the Corriedale ewes deserted their lambs, whilst Shelley (1970) recorded 21% in primiparous and 13% in multiparous Merino ewes. Much lower values usually occur; for example, Arnold and Morgan (1975) had 2% of temporary and 2% of permanent desertions.

There may be genetic differences between flocks in propensity to deser-

Plate 5.6. Ewe deserting her lamb soon after birth.

tion; for example, desertions in the flock of Corriedales observed by Winfield cannot be attributed to other causes. Nutritional status of the ewe can have a strong influence. A.J. Charlick (personal communication, 1974) found that the maternal drive in Merino ewes lambing on pastures deficient in protein was increased significantly when protein supplement was fed. Poor maternal behaviour and a high incidence of desertion have been observed by pastoralists to occur when ewes are in extremely poor condition, lambing under drought conditions. It would seem to be a natural biological mechanism to allow for the survival of the breeding stock in a population.

Primiparous mares frequently exhibit a degree of resentment toward the foal when it contacts the abdomen and udder, and may move away or even kick and bite the foal. Foals may also be prevented from sucking in the first 12 h of life because young mares sometimes appear anxious about losing sight of their foals, turning their heads and bodies around to follow them. This pattern is seen also in ewes and cows.

Chapter 6

THE EFFECTS OF MAN'S HUSBANDRY ON THE BEHAVIOUR AND PRODUCTIVITY OF HIS ANIMALS AND ON THE ENVIRONMENT

ADAPTATION OF BEHAVIOUR TO MAN'S HUSBANDRY

It was pointed out in the introduction to this book that plasticity of behaviour is a major feature of domestic animals. It has been shown in earlier chapters how grazing behaviour is altered to suit the feed conditions available to the animals; changes which allow the animal to sustain its food intake under a wide range of rangeland or pasture conditions. There are, of course, limitations beyond which the various components of grazing behaviour cannot be changed and food intake declines. As was shown, these conditions may arise when feed is very scant, and also when it is very abundant but of low bulk density, a combination which occurs with tropical pastures. Stobbs (1973a, 1974) studied this situation and concluded that dairy cows grazing some tropical swards can have difficulty harvesting sufficient feed to maintain stable production. The reason why the animals have difficulty harvesting enough food is because they are so selective; similar restrictions in intake occur when sheep will spend 12 h a day selecting out small green shoots from a bulk of dry pasture.

Grazing behaviour is adapted to meet quite diverse systems of grazing management, so that food intake and production of dairy cows is similar under continuous grazing, rotational grazing and strip (or break) grazing of pastures (Hodgson, 1933; Waite et al., 1951; Hancock and McMeekan, 1954) even though these different management practices may result in quite different diurnal patterns of activity. The normal grazing period around dawn is foregone under a strip grazing system and transferred to the period when a new strip of pasture is made available (Tayler, 1953; Hancock, 1954a).

Restriction of the time allowed for grazing can be considerably compensated for by more intensive eating during the time allotted. For example, a reduction in the time on the pasture for dairy cows from 1234 to 507 min reduced grazing time by 46 min (Hancock, 1954a). Smith (1961) found that the liveweight gain of cattle on veld in Zambia was similar with a 24 h and 11 h grazing day, but with a 7 h grazing day cattle growth was reduced (a) during the dry season when the veld grazing was deficient in quality, and (b) when there was a quantitative shortage of feed. He found that with restricted time the cattle largely postponed resting and rumination until re-entry into the kraal (or boma).

Other problems of night boma-ing in Africa have been discussed by Goldson (1976). For example, under hot humid conditions, if grazing is allowed only during the day, then quite clearly the animals will not increase their grazing activity unless they are physiologically adapted to the climate.

Under rangeland conditions, the number of and distance between watering points is important, food intake and hence production being reduced when animals are forced to graze on scant feed within range of water whilst abundant feed lies outside this area. Knowing the distances different classes of stock will walk on different types of vegetation is necessary for planning the location of water supplies.

From the information available on mating behaviour it is clear that the ratio of males to females needs to be different in different topographical situations to maximise conception rates in sheep and cattle. More males are needed when paddocks are large, or undulating, or when stock density is low because sexual contact is less easily achieved due to visual barriers and distance. More are also needed with maiden females because they are less active in seeking males and have a short oestrus period.

In chickens and pigs kept under intensive conditions stress from crowding may cause depression in production. This seems unlikely to be a problem with free-ranging domestic animals. Sheep often choose to graze in flocks within which the density is 100 sheep ha^{-1}. Thus, forcing them to graze at this density, as may occur under rotational grazing, is unlikely to cause stress. However, if ewes that are to lamb are grazed at 100 sheep ha^{-1}, they may not be able to isolate themselves at parturition to minimise interference from other pre-parturient ewes. Winfield (1970) found that more lambs of less than 2 h old were separated from their dams when the ewes were at a density of 143 ha^{-1} compared with 14.3 ha^{-1}. Under cold conditions separation at this time could be fatal, and, as was shown, numbers of lambs may die because of interference and lamb stealing.

One of the more important findings that affects sheep production is that only recently shorn ewes will seek shelter, except under very cold conditions. Thus ewes will not find shelter for the sake of their lambs. In Southern Australian conditions, on average, exposure will cause the deaths of 15 out of every 100 lambs born; on some days nearly all newborn lambs will die of exposure. The provision of shelter, i.e. pens, and forcing the ewes to use them, can drastically reduce lamb losses. For example, Egan et al. (1972) in Victoria showed a reduction from 19.0 to 6.3% in lamb deaths in the first 48 h after birth.

At the other end of the spectrum are losses due to heat stress (e.g. Morgan et al., 1972b). The provision of shelter will reduce these, as will the provision of small paddocks where ewes need travel only a short distance to water. The problem is that the day old lamb follows its mother to water and will succumb to heat stress if it walks a kilometer or so.

The seeking of shade in warm and hot weather is characteristic of sheep, cattle and horses. However, little is known of the effects on the performance of livestock forced to live in paddocks without shade.

There is no published information for sheep or horses, but in the hot, humid environment of Louisiana cattle made better liveweight gains when provided with natural shade than with no shade (McDaniel and Roark, 1956). However, artificial shade did not improve liveweight gains. The reason for this was that cattle grazed in the natural shade from scattered trees, but could only shelter in the artificial shade. In other situations artificial shade has resulted in improved liveweight gains (Ittner et al., 1954 in California; Peacock et al., 1965 in Florida; and McIlvain and Shoop, 1971 in Oklahoma). The type of artificial shade is important — hay held between wire being much more effective than galvanised iron. When artificial shade is effective it appears to be so because it decreases the heat load on the animals, allowing a higher food intake.

Disruption of social structure by separating the young from their dams at weaning time does not appear to affect the growth of lambs, but there has been very little research on this topic. Arnold (unpublished) compared the growth of lambs kept with their mothers but receiving no milk because they had been naturally weaned, with that of lambs grazed with alien ewes, and lambs grazed alone. Lamb growth over 6 weeks was 2.1, 3.1 and 2.2 kg, respectively.

Another social disruption, that of mixing with strangers, also appears to have little effect on production. McIlvain and Shoop (1965a) compared weight gains in Hereford steer yearlings over 3 years when rotated monthly between pastures and when rotated monthly with a completely new set of associates. Weekly weight gains were 1.5 and 2.0 kg, respectively. Social dominance has an effect on productivity only when there is competition for a critical resource, such as a food supplement when such is really necessary and is not acting as a substitute for paddock feed. Competition does not usually occur under grazing, although it may be occurring when feed is in short supply since Donnelly et al. (1970) found that Merino sheep gained significantly less weight under such conditions when grazed with sheep of other breeds than when grazed alone.

Flock size had no effect on growth rate of flocks of 3, 9, 27 and 135 sheep. Donnelly et al. also found that rotation from paddock to paddock per se had no effect on growth rate. Thus a change to a similar environment is not disruptive; changes to dissimilar environments have not been studied.

Social organisation may influence animal performance when home ranges are developed, as with sheep on upland areas of Britain. Hunter (1964) showed that the weaning weights of lambs from different home ranges differed significantly.

Size of paddocks has no effect on productivity of sheep or cattle (Elliott, 1966; Southcott et al., 1968) when stocking rate is equal, and so presumably freedom to move around is not important although it is the view of some graziers that sheep 'do better' when given larger paddocks; Elliott found that wool production was equal in sheep grazed in paddocks of 2 and 20 ha. With cattle, Southcott et al. (1968) obtained equal weight gains in paddocks of 8, 16 and 36 ha.

There is relatively little known about grazing behaviour and disease hazards. The fact that horses and cattle do not graze close to defaecation sites at least for several weeks, in some environments, will reduce the risk of reinfection by larvae that hatch from eggs in the faeces and move onto herbage nearby. However, sheep preferentially graze on herbage grown on urine patches (Keogh, 1973a), enhancing the risk of contracting the disease facial eczma in New Zealand. Keogh (1973b) established that herbage on urine patches had a higher spore load of the causative fungus *Pithomyas chartarum*.

EFFECTS OF BEHAVIOUR ON THE ENVIRONMENT

The free-grazing animal can exert a marked influence on its environment as has been amply demonstrated in many parts of the world where overgrazing has occurred. We are not concerned with that here but with the influence of the animals' behaviour on the ecology of the land system in which they graze. Non-random use of space, selective grazing and uneven redistribution of nutrients have marked effects on the vegetation.

In the moorlands and mountains of Britain the pattern of grazing that has existed for centuries has resulted in a stable ecosystem that depends for its stability on that grazing system. The various plant communities and their distribution are dependent, in part, on the soil and topographic situation but also on the sheep. The heavy intensity of grazing on the *Agrostis—Festuca* community brings with it a proportionately high return of nutrients through faeces and urine and faster cycling of nutrients because there is rarely, if ever, a large sink of nutrients in standing biomass as occurs on other types of vegetation, and maintains the community.

In other rangelands where grazing intensity is uneven, similar situations develop. The composition of species on intensively grazed areas, which originally may have been similar to that on low grazing intensity areas, will change. The changes have to be specified for each type of rangeland, but some generalisations are possible. Heavily grazed areas may become completely denuded; this occurs around watering sites in much of Australia's rangeland. The amount and composition of the range may change progressively with distance from the animals' water supply. Under heavy grazing annuals tend to replace perennials, and grasses and other herbaceous species will replace shrubs and trees if these plant types are edible. At the furthest distances from water, shrub and tree vegetation may increase.

In many parts of Australia a major problem is caused by an increase in unpalatable shrubs when herbaceous plants have been intensively grazed. Similar situations occur in North American rangelands. When this occurs the carrying capacity of the range is reduced. A change in animal species may help to alter the composition in the desired direction. Goats, because they browse more than sheep and cattle have been used for this purpose in Africa and North America (Davis et al., 1976; Merrill and Taylor, 1976). However, this practice is not always successful and without strict control goats can convert forest to herbaceous and shrubby vegetation. Examples of this in New Zealand are given by Atkinson (1964).

On pasture lands changes in botanical composition will also occur, with heavily grazed areas more likely to become legume dominant and poorly grazed areas to become grass dominant.

The effects of transfer of nutrients can have a strong localised effect, with quite different botanical composition on and near camping sites. Hilder (1966b) showed that on native pastures of the New England area of New South Wales 80 years of sheep grazing significantly reduced the available phosphorus and potassium in the soil of the major part of many paddocks.

As has been discussed earlier, in small paddocks horses rapidly change a uniform pasture into areas of highly grazed pasture, rank areas reserved for elimination, and bare areas used for rolling (Ödberg, 1975). The result is that renovation is often required every 2—3 years because the carrying capacity of the paddocks is drastically reduced. On large paddocks (>10 ha) only rolling areas are defined. Elimination, whilst not random, is not so confined that a whole section of a paddock is left ungrazed. The effect of paddock size on this type of behaviour needs to be investigated.

Cattle, whilst not developing specific elimination areas even in small paddocks, reject the herbage growing around faeces deposits. Each defaecation by a cow or steer covers an area of about 0.1 m^2 and affects the growth and makes herbage unpalatable, in an area six times the size (Brockington, 1972). Under intensive pasture systems in temperate areas, where a grazing season may be 180 days, up to 45% of the area may be covered with herbage rejected by cattle by the end of the grazing season (Marsh and Campling, 1970). Each urination affects the growth on about 1 m^2, but herbage growth on the sites is preferentially grazed by cattle.

Sheep do not reject herbage around their own faeces, except around sheep camps where it is often very dense, or that around cattle faeces. Therefore combined grazing of sheep and cattle could be used to avoid the wastage described.

The only management practice which ensures a more uniform distribution of nutrients is to have rotational grazing around the paddocks in use. Rotational grazing will not affect the degree of wastage from faeces contamination (Arnold and Holmes, 1958) because only by chance are faeces dropped on sites where faeces already lie.

The tracks made by animals may be striking features of many pastoral

Plate 6.1. Network of sheep tracks 'terracing' a grassy slope.

Plate 6.2. Erosion beginning where sheep tracks cross between ridges.

scenes. The chalk hills of the North and South Downs in Britain that have been grazed by sheep for centuries show a permanent network of tracks (Plate 6.1). In other places the paths across from one ridge to another often join, and on a susceptible soil can be a locus for the beginning of erosion even though overgrazing is not occurring (Plate 6.2).

Improving the use of space, particularly on rangelands, requires recognition of the free-grazing animals' behaviour. Where animals depend on man-made water supplies, these can be located and then manipulated, by turning on or off, to make more uniform use of the vegetation. Even on small paddocks, siting of water away from shade will ensure more even grazing. Other techniques have been used successfully. For example, the use of strips of fertilizer (Smith and Lang, 1958) or reseeding areas high up slopes on hilly or mountain country will attract livestock to the areas and they will utilise intermediate areas as well. The most traditional and effective way of affecting use of space on rangeland is by the use of salt to which cattle are attracted. Jardine and Anderson (1919) presented plans for the distribution of saltings on cattle ranges in the National Forests of the United States. The principle is the same as for siting of water, i.e. salt should be sited at a distance from water, and if possible from shade. Artificial shade, where no natural shade exists, can also be used to manipulate cattle dispersion (McIlvain and Shoop, 1965b).

In many areas of the world herding or shepherding is practised, for two reasons — control of grazing and prevention of theft or loss from predators. But man, it seems, is less effective than the animal itself in minimising damage to the environment, at least where sheep are concerned. Jardine (1910) found that summer mountain range would carry more sheep under free-ranging than under herded conditions because the sheep utilised more country when free-ranging. Jones and Paddock (1966) found in the San Isobel National Forest, Colorado, that trailing damage was reduced and carrying capacity could be doubled, with less erosion damage, when the sheep were allowed to graze freely rather than be herded. Productivity of the sheep is usually similar under the two systems (Strassia et al., 1970).

USE OF DOMESTICATED GRAZING ANIMALS IN MANAGEMENT OF LAND FOR AMENITY USE

The aesthetic value of grazing animals is considerable. To millions of urban dwellers much of the pleasure of the countryside is derived from the sight of animals grazing. A vista of the countryside may have visual appeal for the grandeur of its mountains or the winding of a river across a plain. But to marry the same scene with animals grazing gives extra life to the vista. To others the pastoral scene gives a feeling of tranquility and safety. In the western world urban man is spending more time and has more interest in the countryside, whilst the economic difficulties of ranching or running sheep on small hill farms or on rangeland are forcing the abandonment of marginal agricultural areas. Thiele-Wittig (1974) estimates that in the European Economic Community 10 million hectares of land will become unused for agriculture by 1985. Some of this land may be reafforested but the remainder will need some management to maintain the visual value of the countryside. It may be economic to develop large areas in mountain country with unherded sheep grazing, as occurs in similar areas in Britain. But there is a real need for minimum care systems with little or no animal product being obtained, to preserve these areas of countryside. Such systems are also needed for the maintenance of smaller amenity areas, in parks, on country estates and grassland conservation areas like the Aston Rowant Nature Reserve in England.

The grasslands and forests of the world are grazed by wild ungulates and these animals are often of considerable economic and aesthetic importance to man's recreation needs. In northern Europe, red, roe and fallow deer are the main species. In North America the white-tailed and black-tailed deer and pronghorn antelope and elk are of major importance and often cohabit the rangelands with cattle or sheep. In Australia their counterparts are species of kangaroos and wallabies. Almost all of these species are difficult to domesticate in large numbers or to manage easily.

Sheep are the obvious solution for such countryside but breeds are needed that will adapt to a partial or complete reversion to their original natural state. One important characteristic would be the ability to shed wool. There are examples around the world where sheep have become feral after being abandoned by man, and have successfully maintained a population. Feral goats exist in many parts of the world but their destructiveness of trees and shrubs makes them of low value for the purpose in mind.

On the islands of Soay and Hirta, in the St. Kilda group in the outer Hebrides, the Soay sheep were introduced when the last farmers left in 1930. This primitive breed of sheep has survived and has been studied extensively. Sheep from the island have been successfully used in parks in England (Plate 6.3). On an adjacent island, Boreray, the original sheep were of the Lewis strain of Scottish Blackface and these were abandoned when the island was deserted in 1930, but in 1956 a population of 365 existed on the 77 ha island. Fleece shedding, which is natural in the Soay, can occur in the Blackface, possibly because of a nutritional break in the wool in late winter.

A preliminary study has been made of the Soay in managing amenity grassland in the Chiltern Hills in England (Large and Spedding, 1976). These workers attempted to graze Soay ewes year round and cross them with a Down breed of ram. The aim was to obtain some meat output to offset the cost of managing the sheep. Some nutritional problems arose and further work is needed.

On mountainous Campbell Island, south of New Zealand, sheep have been

Plate 6.3. Soay sheep grazing in parkland in England.

Plate 6.4. Feral sheep on Campbell Island, New Zealand. (Photography courtesy of Mr. R.H. Taylor, DSIR, Nelson, New Zealand.)

feral for 70 years. They are a half-breed Merino longwool cross (Wilson and Orwin, 1964) — Plate 6.4 — and exist in substantial numbers, more than 2000 in 1969 (Taylor et al., 1970).

Plate 6.5. Natural flocks of Dorset Horn and Southdown sheep in Australia.

Plate 6.6. Jacob sheep.

Plate 6.7. A mixed flock of black and white sheep grazing in parkland in England.

Plate 6.8. Wiltshire Horn sheep.

In Australia, sheep of four breeds — Merino, Dorset Horn, Cheviot and Southdown — have been studied as natural populations (Plate 6.5) for fifteen years (Arnold, unpublished). There is a minimum of management; sheep are shorn and treated for flystrike (a major problem in Australia but not in the cold climates of St. Kilda or Campbell Island). The populations have resulted in strains of Merinos and Dorset Horn sheep that are much hardier than husbanded flocks. The flocks develop social structures similar to those in the primitive Soay sheep (Grubb and Jewell, 1966). Reproductive rate is high in the Merinos and Dorset Horns which lamb at about 6.5 month intervals. The Southdown and Cheviot ewes rarely lamb outside a short period in winter and spring.

There is no doubt that many breeds of sheep can adapt, behaviourally, back to the natural state and could be managed in this way to maintain the countryside. Variety could be developed using various breeds, indeed some, like the Jacob (Plate 6.6) and black sheep (Plate 6.7) may be kept primarily for aesthetic reasons. Besides wool shedding (e.g. as in the Wiltshire Horn, Plate 6.8), other attributes needed are being attractive to look at, hardy and independent, fleet of foot and with horns for defence.

Fallow and roe deer inhabit the parks of several stately homes in Britain and might well be used for maintaining areas of countryside where closer control, mainly fences, is possible.

REFERENCES

Alexander, G., 1958. Behaviour of nowly born lambs. Proc. Aust. Soc. Anim. Prod., 2: 123—125.

Alexander, G., 1960. Maternal behaviour in the Merino ewe. Proc. Aust. Soc. Anim. Prod., 3: 105—114.

Alexander, G., 1977. Role of auditory and visual cues in mutual recognition between ewes and lambs in Merino sheep. Appl. Anim. Ethol., 3: 65—81.

Alexander, G. and Shillito, E.E., 1977a. The importance of odour, appearance and voice in maternal recognition of the young in Merino sheep (*Ovis aries*). Appl. Anim. Ethol., 3: 127—135.

Alexander, G. and Shillito, E.E., 1977b. Importance of visual clues from various body regions in maternal recognition of the young in Merino sheep (*Ovis aries*). Appl. Anim. Ethol., 3: 137—143.

Alexander, G. and Williams, D., 1964. Maternal facilitation of sucking drive in new born lambs. Science, 146: 665—666.

Alexander, G. and Williams, D., 1966. Teat seeking activity in new born lambs: the effects of cold. J. Agric. Sci. Camb., 67: 181—189.

Alexander, G., Mann, T., Mulhearn, C.J., Rowley, I.C.R., Williams, D. and Winn, D., 1967. Activities of foxes and crows in a flock of lambing ewes. Aust. J. Exp. Agric. Anim. Husb., 7: 329—336.

Allden, W.G. and Whittaker, I.A. McD., 1970. The determinants of herbage intake by grazing sheep. The interrelationships of factors influencing herbage intake and availability. Aust. J. Agric. Res., 21: 755—766.

Allison, A.J. and Davis, G.H., 1976a. Studies of mating behaviour and fertility of Merino ewes. I. Effects of number of ewes joined per ram, age of ewe, and paddock size. N.Z.J. Exp. Agric., 4: 259—267.

Allison, A.J. and Davis, G.H., 1976b. Studies of mating behaviour and fertility of Merino ewes. II. Effects of age of ewe, live weight, and paddock size on duration of oestrus and ram-seeking activity. N.Z.J. Exp. Agric., 4: 269—274.

Archer, M., 1972. Pasture management. Br. Racehorse, Dec. 1972, pp. 609—611.

Arnold, G.W., 1960a. The effect of the quantity and quality of pasture available to sheep on their grazing behaviour. Aust. J. Agric. Res., 11: 1034—1043.

Arnold, G.W., 1960b. Selective grazing by sheep of two forage species at different stages of growth. Aust. J. Agric. Res., 11: 1026—1033.

Arnold, G.W., 1962. The influence of several factors in determining the grazing behaviour of Border Leicester × Merino sheep. J. Br. Grassl. Soc., 17: 41—51.

Arnold, G.W., 1964a. Factors within plant associations affecting the behaviour and performance of grazing animals. In: D.J. Crisp (Editor), Grazing in Terrestrial and Marine Environments. Blackwell, Oxford, pp. 133—154.

Arnold, G.W., 1964b. Some principles in the investigation of selective grazing. Proc. Aust. Soc. Anim. Prod., 5: 258—271.

Arnold, G.W., 1964c. Grazing experience in early life and subsequent performance in sheep. Field Stn Rec., Div. Plant Ind., CSIRO (Aust.), 3: 13—20.

Arnold, G.W., 1964d. Supplementation of *Phalaris tuberosa—Trifolium subterraneum* pasture in summer with urea and molasses. Field Stn Rec., Div. Plant Ind., CSIRO (Aust.), 3: 37—44.

Arnold, G.W., 1964e. Responses of lambs to differing pasture conditions. Proc. Aust. Soc. Anim. Prod., 5: 275—279.

Arnold, G.W., 1966a. The special senses in grazing animals. I. Sight and dietary habits in sheep. Aust. J. Agric. Res., 17: 521—529.

Arnold, G.W., 1966b. The special senses in grazing animals. II. Smell, taste, and touch and dietary habits in sheep. Aust. J. Agric. Res., 17: 531—542.

Arnold, G.W., 1966c. The use of in vitro methods associated with sampling techniques to measure intake and digestibility under grazing conditions. In: O. Paladines (Editor), Symposium 'Determination of the Nutritive Value of Forages. In Vitro Methods'. Centro De Investigacion Y Enseñanza Para La Zona Templada, Instituto Interamericano De Ciencias Agricolas De La Oea, La Estanzuela, Colonia, Uruguay, October 1966, pp. 63—97.

Arnold, G.W., 1970. Regulation of food intake in grazing ruminants. In: D. Cuthbertson (Editor), Proc. 3rd Int. Symposium 'The Physiology of Digestion and Metabolism in the Ruminant'. Cambridge, Great Britain, pp. 264—276.

Arnold, G.W., 1975. Herbage intake and grazing behaviour in ewes of four breeds at different physiological states. Aust. J. Agric. Res., 26: 1017—1024.
Arnold, G.W., 1976. A note on changes in ingestive behaviour of sheep following shearing. Appl. Anim. Ethol., 2: 175—179.
Arnold, G.W., 1977. An analysis of spatial leadership in a small field in a small flock of sheep. Appl. Anim. Ethol., 3: 263—270.
Arnold, G.W. and Baas Becking, H.G., 1964. Dominance ranking amongst four rams and its relationship to leadership and mating behaviour. Film, produced by CSIRO Division of Plant Industry, Canberra, Australia.
Arnold, G.W. and Birrell, H., 1977. Food intake and grazing behaviour of sheep varying in body condition. Anim. Prod., 24: 343—353.
Arnold, G.W. and Dudzinski, M.L., 1966. The behavioural responses controlling the food intake of grazing sheep. Proc. XI Int. Grassl. Congr., Finland. Valtioneuvoston Kirjapaino, Helsinki, pp. 367—370.
Arnold, G.W. and Dudzinski, M.L., 1967a. Studies on the diet of the grazing animal. II. The effect of physiological status in ewes and pasture availability on herbage intake. Aust. J. Agric. Res., 18: 349—359.
Arnold, G.W. and Dudzinski, M.L., 1967b. Studies on the diet of the grazing animal. III. The effect of pasture species and pasture structure on the herbage intake of sheep. Aust. J. Agric. Res., 18: 657—666.
Arnold, G.W. and Dudzinski, M.L., 1969. The effect of pasture density and structure on what the grazing animal eats and animal productivity. In: B.J.F. James (Editor), Intensive Utilisation of Pastures. Angus and Robertson, Sydney, N.S.W., pp. 42—48.
Arnold, G.W. and Hill, J.L., 1972. Chemical factors affecting selection of food plants by ruminants. In: J.B. Harborne (Editor), Phytochemical Ecology. Academic Press, London, pp. 71—101.
Arnold, G.W. and Holmes, W., 1958. Studies in grazing management. VII. The influence of strip grazing versus controlled free grazing on milk yield, milk composition and pasture utilisation. J. Agric. Sci. Camb., 51: 248—256.
Arnold, G.W. and Maller, R.A., 1974. Some aspects of competition between sheep for supplementary feed. Anim. Prod., 19: 309—319.
Arnold, G.W. and Maller, R.A., 1977. Effects of nutritional experience in early and adult life on the performance and dietary habits of sheep. Appl. Anim. Ethol., 3: 5—26.
Arnold, G.W. and Morgan, P.D., 1975. Behaviour of the ewe and lamb at lambing and its relationship to lamb mortality. Appl. Anim. Ethol., 2: 25—46.
Arnold, G.W. and Pahl, P.J., 1967. Sub-grouping in sheep flocks. Proc. Ecol. Soc. Aust., 2: 183—189.
Arnold, G.W. and Pahl, P.J., 1974. Some aspects of social behaviour in domestic sheep. Anim. Behav., 22: 592—600.
Arnold, G.W., McManus, W.R., Bush, I.G. and Ball, J., 1964. The use of sheep fitted with oesophageal fistulas to measure diet quality. Aust. J. Exp. Agric. Anim. Husb., 4: 71—79.
Arnold, G.W., Ball, J., McManus, W.R. and Bush, I.G., 1966. Studies on the diet of the grazing animal. I. Seasonal changes in the diet of sheep grazing on pastures of different availability and composition. Aust. J. Agric. Res., 17: 543—556.
Arnold, G.W., Boundy, C.A.P., Morgan, P.D. and Bartle, G., 1975. The role of sight and hearing in the lamb in the location and discrimination between ewes. Appl. Anim. Ethol., 1: 167—176.
Arnold, G.W., Wallace, S.R. and Maller, R.A., 1979. Some factors involved in natural weaning processes in sheep. Appl. Anim. Ethol., 5 (in press).
Arthur, G.H., 1961. Some observations on the behaviour of parturient farm animals with particular reference to cattle. Vet. Rev., 12: 75—84.
Ashton, W.M. and Jones, E., 1959. Coumarin and related components in sweet vernal. J. Br. Grassl. Soc., 14: 47—54.
Askins, G.D. and Turner, E.E., 1972. A behavioural study of angora goats on West Texas range. J. Range Manage., 25: 82—87.
Atkinson, I.A.E., 1964. Relations between feral goats and vegetation in New Zealand. Proc. N.Z. Ecol. Soc., 11: 39—44.
Bailey, P.J., Bishop, A.H. and Boord, C.T., 1974. Grazing behaviour of steers. Proc. Aust. Soc. Anim. Prod., 10: 303—306.
Barnes, R.F., Martin, G.C. and Simons, A.B., 1969. Indole alkaloid derivatives in *Phalaris arundinacea* L. clones of varying palatability. Agron. Abstr., p. 57.
Barty, K., 1974. Observations and procedures at foaling on a thoroughbred stud. Aust. Vet. J., 50: 553—557.
Baskin, L.M., 1975. Distribution of animals on a pasture as a function of the group behaviour. S. Kh. Biol., 10: 407—411.
Bedell, T.E., 1968. Seasonal forage preferences of grazing cattle and sheep in Western Oregon. J. Range Manage., 21: 291—297.
Bedell, T.E., 1971. Nutritive value of forage and diets of sheep and cattle from Oregon subclover—grass mixtures. J. Range Manage., 24: 125—133.

Beerwinkle, L.G., 1974. Heat detection programs and techniques. Proc. 8th Annu. Conf. N.A.A.B., Denver: Beef Artificial Insemination. N.A.A.B., Columbia, Miss., pp. 24—29.

Beilharz, R.G. and Mylrea, P.J., 1963. Social position and movement orders of dairy heifers. Anim. Behav., 11: 529—533.

Bennie, J.G.S., 1956. Grazing behaviour tests with Mauritius Creole cows. Emp. J. Exp. Agric., 24: 312—316.

Bishop, E.J.B., 1965. Shelter for your stock. Farming S. Afr., 41: 43—45.

Bland, B.F. and Dent, J.W., 1964. Animal preference in relation to the chemical composition and digestibility of varieties of cocksfoot. J. Br. Grassl. Soc., 19: 303—315.

Blauvelt, H., Richmond, J.B. and Moore, A.V., 1960. The development of contact between mother and offspring in ungulates (domestic sheep and goats). Bull. Ecol. Soc. Am., 41: 91.

Blockey, M.A. de B., 1976. Sexual behaviour of bulls at pasture: a review. Theriogenlogy, 6: 387—391.

Blockey, M.A. de B. and Cumming, I.A., 1970. Mating behaviour of Merino ewes. Proc. Aust. Soc. Anim. Prod., 8: 344—347.

Bond, J., Carlson, G.E. and Curry, W.A., 1967. Social cohesion of steers and sheep as a possible variable in grazing studies. Agron. J., 59: 481—482.

Bonsma, J.C. and Le Roux, J.D., 1953. Influence of environment on the grazing habits of cattle. Farming S. Afr., 28: 43—46.

Bouissou, M.-F., 1968. Effect de l'ablation des bulbes olfactifs sur la reconnaissance du jeune par sa mere chez les ovins. Cent. Rech. Vet. Zootech., 37: 77—83.

Bourke, M.E., 1967. A study of mating behaviour of Merino rams. Aust. J. Exp. Agric. Anim. Husb., 7: 203—205.

Bowns, J.E., 1971. Sheep behaviour under unherded conditions on mountain summer ranges. J. Range Manage., 24: 105—109.

Box, T.W., Brown, G. and Liles, J., 1965. Influence of winter supplemental feeding of cottonseed cake on activities of beef cows. J. Range Manage., 18: 124—126.

Bredon, R.M., Torell, D.T. and Marshall, B., 1967. Measurement of selective grazing of tropical pastures using oesophageal fistulated steers. J. Range Manage., 29: 317—320.

Bremner, K.J., 1975. Social interactions among dairy cows during herd formation in spring. Proc. N.Z. Soc. Anim. Prod., 35: 231—237.

Brockington, N.R., 1972. A mathematical model of pasture contamination by grazing cattle and the effects on herbage intake. J. Agric. Sci. Camb., 79: 249—257.

Broster, W.H., Tuck, V.J. and Balch, C.C., 1963. Effects of rationing grass on the growth rate of dairy heifers and on output per acre, with a note on its significance in experimental design. J. Agric. Sci. Camb., 69: 371—380.

Brown, J.A.M., 1961. Evaluation of certain morphological and chemical characteristics in relation to palatability in reed canary grass (*Phalaris arundinacea* L.). Diss. Abstr., 22: 373.

Brown, T.H., 1964. The early weaning of lambs. J. Agric. Sci. Camb., 63: 191—204.

Bruemmer, F., 1967. The wild horses of Sable Island. Anim. Can., 10: 14—17.

Brumby, P.J., 1959. The grazing behaviour of dairy cattle in relation to milk production, live weight, and pasture intake. N.Z. J. Agric. Res., 2: 797—807.

Bulbenik, A.B., 1961. Vierundzwanzigstunden-Rhythmus des Przewalski-Pferdes (*Equus przewalski* Pol 1881) während der Laktation und beginnender Brunst. Equus, 1: 122—140.

Bush, L.P., Boling, J.A., Allen, G. and Bucknew, R.C., 1972. Inhibitory effects of perloline to rumen fermentation in vitro. Crop Sci., 12: 277—279.

Cahill, L.P., Blockey, M.A. de B. and Parr, R.A., 1975. Effects of mating behaviour and ram libido on the fertility of young ewes. Aust. J. Exp. Agric. Anim. Husb., 15: 337—341.

Castle, K.E., Foot, A.S. and Halley, R.J., 1950. Some observations on the behaviour of dairy cattle with particular reference to grazing. J. Dairy Res., 17: 215—230.

Chambers, D.T., 1959. Grazing behaviour of calves reared at pastures. J. Agric. Sci. Camb., 53: 417—424.

Charcon, E. and Stobbs, T.H., 1976. Influence of progressive defoliation of a grass sward on the eating behaviour of cattle. Aust. J. Agric. Res., 27: 709—727.

Coleman, S.W., Borth, K.M., McLaren, J.B., High, T.W., Wang, H.C. and Hobbs, C.S., 1971. Composition of Diets Selected by Grazing Beef Cattle Determined after Adjusting for Oesophageal Fistula Effects. Bull. Univ. Tenn. Agric. Exp. Stn, 483, 22 pp.

Collery, L., 1974. Observations of equine animals under farm and feral conditions. Equine Vet. J., 6: 170—173.

Collias, N.E., 1956. The analysis of socialization in sheep and goats. Ecology, 37: 228—239.

Compton, T.L. and Brundage, A.L., 1971. Cattle behavior on subalpine range in southcentral Alaska. J. Anim. Sci., 32: 339—342.

Cook, C.W., 1959. The effect of site on the palatibility and nutritive content of seeded wheat grasses. J. Range Manage., 12: 289—292.

Cook, C.W., 1966. Factors affecting utilization of mountain slopes by cattle. J. Range Manage., 19: 200—204.

Cook, C.W. and Harris, L.E., 1950. The Nutritive Content of the Grazing Sheep's Diet on Summer and Winter Range in Utah. Bull. Utah Agric. Exp. Stn, 342, 66 pp.
Cook, C.W., Stoddart, L.A. and Harris, L.E., 1956. Comparative Nutritive Value and Palatability of some Introduced and Native Forage Plants for Spring and Summer Grazing. Bull. Utah Agric. Exp. Stn. 385, 39 pp.
Cook, C.W., Harris, L.E. and Young, M.C., 1967. Botanical and nutritive content of diets of cattle and sheep under single and common use on mountain range. J. Anim. Sci., 26: 1169—1174.
Coop, I.E. and Drake, J.H., 1949. Shearing ewes before lambing. Proc. N.Z. Soc. Anim. Prod., 9: 122—126.
Cory, V.L., 1927. Activities of Livestock on the Range. Bull. Texas Agric. Exp. Stn, 367, 47 pp.
Cowan, R.T. and O'Grady, P., 1976. Effect of presentation yield of a tropical grass—legume pasture on grazing time and milk yield of Friesian cows. Trop. Grassl., 10: 213—218.
Cowlishaw, S.J. and Alder, F.E., 1960. The grazing preferences of cattle and sheep. J. Agric. Sci. Camb., 54: 257—265.
Cresswell, E., 1960. Ranging behaviour studies with Romney Marsh and Cheviot sheep in New Zealand. Anim. Behav., 8: 1—2.
Crofton, H.D., 1958. Nematode parasite populations in sheep on lowland farms. VI. Sheep behaviour and nematode infections. Parasitology, 48: 251—260.
Croker, K.P. and Lindsay, D.R., 1972. A study of the mating behaviour of rams when joined at different proportions. Aust. J. Exp. Agric. Anim. Husb., 12: 13—18.
Crook, I., 1969. Feral goats of North Wales. Animals (London), 12: 13—15.
Culley, M.J., 1938. Grazing habits of range cattle. J. For., 36: 715—717.
Davies, H.L., 1964. Lamb losses in South Western Australia. Proc. Aust. Soc. Anim. Prod., 5: 107—112.
Davis, G.G., Bartel, L.E. and Cook, C.W., 1976. Control of Gambel oak sprouts. J. Range Manage., 28: 216—218.
Davis, W.B., 1938. Summer activity of mountain sheep on Mt. Washburn, Yellowstone National Park. J. Mammal., 19: 88—94.
Daws, G.T. and Squires, V.R., 1974. Observations on the effects of temperature and distance to water on the behaviour of Merino and Border Leicester sheep. J. Agric. Sci. Camb., 82: 383—390.
De Alba, J. and Asdell, S.A., 1946. Estrous behaviour and hormones in the cow. J. Comp. Physiol., 39: 119—123.
Dennis, S.M., 1969. Predators and perinatal mortality of lambs in Western Australia. Aust. Vet. J., 45: 6—9.
Denton, D.A. and Sabine, J.R., 1963. The behaviour of Na deficient sheep. Behaviour, 20: 264—276.
De Vuyst, A., Thines, G., Heriet, L. and Soffie, M., 1964. Influence des stimulations auditives sur le comportement sexuel du taureau. Experientia, 20: 648—653.
Donald, A.D. and Leslie, R.T., 1969. Population studies on the infective stage of some nematode parasites of sheep. II. The distribution of faecal deposits on fields grazed by sheep. Parasitology, 59: 141—157.
Donaldson, L.E., 1962. Competition between Brahman cross and Hereford bulls in mating. Aust. Vet. J., 38: 520.
Donelly, E.D., 1954. Some factors that affect palatibility in *Sericea lespedeza* (*L. cuneata*). Agron. J., 46: 96—97.
Donnelly, J.R., Axelsen, A. and Morley, F.H.W., 1970. Effect of flock size and grazing management on sheep production. Aust. J. Exp. Agric. Anim. Husb., 10: 271—278.
Dove, W.F., 1935. A study of individuality in the nutritive instincts and of the causes of and effects of variations in the selection of food. Am. Nat., 69: 469—600.
Dudzinski, M.L. and Arnold, G.W., 1967. Aerial photography and statistical analysis for studying behaviour patterns of grazing animals. J. Range Manage., 29: 77—83.
Dudzinski, M.L. and Arnold, G.W., 1973. Comparisons of diets of sheep and cattle grazing together on sown pastures on the southern tablelands of New South Wales by principal components analysis. Aust. J. Agric. Res., 24: 899—912.
Dudzinski, M.L. and Arnold, G.W., 1979. Factors influencing the grazing behaviour of sheep in a Mediterranean climate in summer. Appl. Anim. Ethol., 5 (submitted).
Dudzinski, M.L., Pahl, P.J. and Arnold, G.W., 1969. Quantitative assessment of grazing behaviour of sheep in arid areas. J. Range Manage., 22: 230—235.
Dudzinski, M.L., Schuh, H-J., Wilcox, D.G., Gardiner, H.G. and Morrissey, J.G., 1978. Statistical and probabilistic estimators of forage conditions from grazing behaviour of Merino sheep in a semi-arid environment. Appl. Anim. Ethol., 4: (in press).
Dukes, H.H., 1955. The Physiology of Domestic Animals. 7th Edition. Comstock, Ithaca, N.Y., 1020 pp.
Dwyer, D.D., 1961. Activities and grazing preferences of cows with calves in Northern Osage County, Oklahoma. Bull. Okla Agric. Exp. Stn, B-588, 61 pp.
Egan, J.K., McLaughlin J.W., Thompson, R.L. and McIntyre, J.S., 1972. The importance of shelter in reducing neonatal lamb deaths. Aust. J. Exp. Agric. Anim. Husb., 12: 470—472.

Ehrenreich, J.H. and Bjugstad, A.J., 1966. Cattle grazing time is related to temperature and humidity. J. Range Manage., 19: 141—142.
Elliott, N.M., 1966. The effect of paddock size on annual production. Proc. Aust. Soc. Anim. Prod., 6: 177—178.
Evans, P.S., 1964. A study of leaf strength in four ryegrass varieties. N.Z. J. Agric. Res., 7: 508—513.
Ewbank, R., 1963. Predicting the time of parturition in the normal cow: a study of the precalving temperature in relation to the external signs of imminent calving. Vet. Rec., 75: 367—371.
Ewbank, R., 1964. Observations on the suckling habits of twin lambs. Anim. Behav., 12: 34—37.
Ewbank, R., 1967a. Behaviour of twin cattle. J. Dairy Sci., 50: 1510—1512.
Ewbank, R., 1967b. Nursing and suckling behaviour amongst Clun Forest ewes and lambs. Anim. Behav., 15: 251—258.
Ewbank, R. and Mason, A.C., 1967. A note on the suckling behaviour of twin lambs reared as singles. Anim. Prod., 9: 417—420.
Fayre, Y., 1976. Amenagément et utilisation du domaine pastoral. Bull. Tech. Inf., 314/315: 661—670.
Firbank, T., 1946. I Bought a Mountain. Readers' Union, George G. Harrap, London, 255 pp.
Fletcher, I.C. and Nicolson, A., 1976. Group structures within a flock of Merino ewes and rams on a large paddock in semi-arid pastoral country. Proc. Aust. Soc. Anim. Prod., 11: 145—148.
Fowler, D.G., 1975. Mating activity and its relationship to reproductive performance in Merino sheep. Appl. Anim. Ethol., 1: 357—368.
Francis, C.M., 1973. The influence of isoflavone glycosides on the taste of subterranean clover leaves. J. Sci. Food Agric., 24: 1235—1240.
Fraser, A., 1937. Sheep Farming. Crosby Lockwood, London, 150 pp.
Fraser, A.F., 1968. Reproductive Behaviour in Ungulates. Academic Press, London, 202 pp.
Fraser, A.H.H., 1926. Chain instincts in lambing sheep. J. Psychol., 16: 311—313.
Galt, H.D., Theurer, B., Ehrenreich, J.H., Hale, W.H. and Martin, S.C., 1969. Botanical composition of diet of steers grazing a desert grassland range. J. Range Manage., 22: 14—19.
Gangstadt, E.O., 1964. Physical and chemical composition of grass sorghum as related to palatibility. Crop Sci., 4: 269—273.
Gardiner, H.G., Wilcox, D.G. and Morrisey, J.G., 1978. The influence of liveweight on group size and mating efficiency in a ewe flock in an arid grazing environment. Aust. Rangel. J., 1: 112—116.
Gary, L.A., Sherritt, G.W. and Hale, E.B., 1970. Behaviour of Charolais cattle on pasture. J. Anim. Sci., 30: 203—206.
Gauthier-Pilters, H., 1974. The behaviour and ecology of camels in the Sahara, with special reference to nomadism and water management. In: V. Geist and F. Walther (Editors), The Behaviour of Ungulates and its Relation to Management. IUCN Publ. New Series No. 24, Morges, Switzerland, pp. 542—551.
Geist, V., 1969. *Ovis canadensis* (Bovidae) rutting behaviour. Encyclopaedic Cinematographica E 1335, pp. 537—545, Göttingen.
Geist, V., 1971. Mountain Sheep. A Study in Behaviour and Evolution. Univ. Chicago Press, Chicago, Ill., 383 pp.
George, J.M., 1969. Variation in the time of parturition of Merino and Dorset Horn ewes. J. Agric. Sci. Camb., 73: 295—299.
George, J.M. and Barger, I.A., 1974. Observations on bovine parturition. Proc. Aust. Soc. Anim. Prod., 10: 314—317.
Gladstones, J.S., 1970. Lupins as crop plants. Field Crop Abstr., 23: 123—148.
Goatcher, W.D. and Church, D.C., 1970. Review of some nutritional aspects of the sense of taste. J. Anim. Sci., 31: 973—981.
Goldson, J.R., 1976. The Practice of Night Kraaling of Dairy Cattle on the Smallholder Farm. Preview of Literature and Proposals for a Research Programme. Tech. Rep. Min. Agric. Kenya Past. Res. Project, 8, 39 pp.
Gonzales, M.H., 1964. Patterns of Livestock Behavior and Forage Utilisation as Influenced by Environmental Factors on a Summer Mountain Range. Ph. D. Thesis, Utah State Univ., Logan, Utah, 115 pp.
Gordon, J.G. and Tribe, D.E., 1951. The self selection of diet by pregnant ewes. J. Agric. Sci., Camb., 41: 187—190.
Gordon, J.G., Tribe, D.E. and Graham, T.C., 1954. The feeding behaviour of phosphorus deficient cattle and sheep. Br. J. Anim. Behav., 2: 72—74.
Gordon, W.C. and Henderson, J.H.M., 1951. The alkaloidal content of blue lupins (*Lupinus angustifolius* (L.)) and its toxicity on small laboratory animals. J. Agric. Sci. Camb., 41: 141—145.
Green, L.R., Wagnon, K.A. and Bentley, J.R., 1958. Diet and grazing habits of steers on foothill range fertilized with sulfur. J. Range Manage., 11: 221—227.
Greenhalgh, J.F.D., Reid, G.W., Aitken, J.N. and Florence, E., 1966. Effects of grazing intensity on herbage consumption and animal production. I. Short term effects in strip grazed dairy cows. J. Agric. Sci. Camb., 67: 13—24.

Griffiths, J.G., 1966. The identification of topographic shelter in hill areas and its use by grazing sheep. Proc. 9th Int. Congr. Anim. Prod., Edinburgh, August 1966. Oliver and Boyd, Edinburgh, p. 67.

Griffiths, J.G., 1967. Behavioural and physiological responses of sheep to climatic exposure. Hill Farming Res. Org., Fourth Rep. 1964—67, pp. 85—90.

Griffiths, J.G., 1970. A note on patterns of behaviour and grazing in hill sheep. Anim. Prod., 12: 521—524.

Grubb, P., 1974. Mating activity and the social significance of rams in a feral sheep community. In: V. Geister and F. Walther (Editors), The Behaviour of Ungulates and its Relation to Management. IUCN Publ. New Series No. 24, Morges, Switzerland.

Grubb, P. and Jewell, P.A., 1966. Social grouping and home range in feral Soay sheep. Symp. Zool. Soc. Lond., No. 18, pp. 179—210.

Grubb, P. and Jewell, P.A., 1973. The rut and the occurrence of oestrus in the Soay sheep on St. Kilda. J. Reprod. Fertil. Suppl., 19: 491—502.

Grzimek, B., 1944. Pecking order experiments with horses. Z. Tierpsychol., 6: 455—464.

Gunn, R.G., 1968. A note on difficult birth in Scottish hill flocks. Anim. Prod., 10: 213—215.

Hafez, E.S.E., 1951. Mating behaviour in sheep. Nature (London), 1967: 777—778.

Hafez, E.S.E., 1959. Reproductive capacity of farm animals in relation to climate and nutrition. J. Am. Vet. Med. Assoc., 135: 606—614.

Hafez, E.S.E. (Editor), 1969. The Behaviour of Domestic Animals. Ballière, Tindall and Cassell, London, 647 pp. (2nd Edition).

Hafez, E.S.E. and Lindsay, D.R., 1965. Behavioural responses in farm animals and their relevance to research techniques. Anim. Breed. Abstr., 33: 1—16.

Hafez, E.S.E. and Scott, J.P., 1962. The behaviour of sheep and goats. In: E.S.E. Hafez (Editor), The Behaviour of Domestic Animals. Ballière, Tindall and Cox, London. (1st Edition)

Hafez, E.S.E., Schein, M.W. and Ewbank, R., 1969. The behaviour of cattle. In: E.S.E. Hafez (Editor), The Behaviour of Domestic Animals. Ballière, Tindall and Cassell, London, pp. 233—295.

Hale, E.B., 1966. Visual stimuli and reproductive behaviour in bulls. J. Anim. Sci., 25: 36—48.

Hall, R., 1972. Wild Horse: Biology for Management. USDI Bur. Land Manage., Billings District, Unit Resource Analysis, 117 pp.

Hammond, J., 1944. In a discussion on "The importance of prosecuting the study of animal behaviour and suggestions for its development as a field of veterinary medicine" by J.T. Edwards. Proc. Br. Soc. Anim. Prod., 1944, p. 103.

Hancock, J., 1950. Studies in monozygous cattle twins. 4. Uniformity trial: grazing behaviour. N.Z. J. Sci. Technol., Sect. A., 32: 22—54.

Hancock, J., 1952. Grazing behaviour of identical twins in relation to pasture type, intake and production of dairy cattle. Emp. J. Exp. Agric., 18: 1399—1407.

Hancock, J., 1954a. Studies on grazing behaviour in relation to grassland management. I. Variations in grazing habits of dairy cattle. J. Agric. Sci. Camb., 44: 420—429.

Hancock, J., 1954b. Studies of grazing behaviour in relation to grassland management. II. Bloat in relation to grazing behaviour. J. Agric. Sci. Camb., 44: 80—95.

Hancock, J. and McArthur, A.T.G., 1951. Tips on cow management arising from grazing behaviour studies. Proc. Ruakura Fmrs Conf., Hamilton, N.Z. 1951, pp. 32—38.

Hancock, J. and McMeekan, C.P., 1954. Studies of grazing behaviour in relation to grassland management. III. Rotational compared with continuous grazing. J. Agric. Sci. Camb., 45: 96—103.

Hardison, W.A., Reid, J.T., Martin, C.M. and Wodfolk, P.G., 1954. Degree of herbage selection by grazing cattle. J. Dairy Sci., 37: 89—101.

Hardison, W.A., Fisher, H.L., Graf, G.C. and Thompson, N.R., 1956. Some observations on the behaviour of grazing lactating cows. J. Dairy Sci., 39: 1735—1741.

Harker, K.W., Taylor, J.I. and Hollinson, D.H.L., 1954. Studies on the habits of Zebu cattle. I. Preliminary observations on grazing habits. J. Agric. Sci. Camb., 44: 193—198.

Harker, K.W., Rollinson, D.H.L., Taylor, J.I., Gourlay, R.N. and Nunn, W.R., 1961. Studies on the habits of Zebu cattle. VI. The results on different pastures. J. Agric. Sci. Camb., 56: 137—141.

Herbel, C.H. and Nelson, A.B., 1966. Species preference of Hereford and Santa Gertrudis cattle on a southern New Mexico range. J. Range Manage., 19: 177—181.

Hersher, L.J., Richmond, B. and Moore, A.U., 1963. Modifiability of the critical period for the development of maternal behavior in sheep and goats. Behaviour, 20: 311—320.

Hilder, E.J., 1966a. Distribution of excreta by sheep at pasture. Proc. 8th Int. Grassl. Congr., pp. 977—981.

Hilder, E.J., 1966b. Rate of turnover of elements in soils: the effects of stocking rate. Wool Technol. Sheep Breed., 13: 11—16.

Hodder, R.M. and Low, W.A., 1978. Grazing distribution of free-ranging cattle at three sites in the Alice Springs district, Central Australia. Aust. Rangel. J., 1: 95—106.

Hodge, R.W. and Doyle, J.J., 1967. Diet selected by lambs and yearling sheep grazing on annual and perennial pastures in southern Victoria. Aust. J. Exp. Agric. Anim. Husb., 7: 141—143.

Hodgson, J. and Wilkinson, J.M., 1967. The relationship between liveweight and herbage intake in grazing cattle. Anim. Prod., 9: 365—376.

Hodgson, J. and Wilkinson, J.M., 1968. The influence of the quantity of herbage offered and its digestibility on the amount eaten by grazing cattle. J. Br. Grassl. Soc., 23: 75—80.
Hodgson, R.E., 1933. Influence of pasture management upon the grazing habits of dairy cattle. J. Agric. Res., 47: 417—424.
Holder, J.M., 1960. Observations on the grazing behaviour of lactating dairy cattle in a sub-tropical environment. J. Agric. Sci. Camb., 55: 261—267.
Holder, J.M., 1962. Supplementary feeding of grazing sheep — its effect on pasture intake. Proc. Aust. Soc. Anim. Prod., 4: 154—159.
Honkanen, E., Molsio, T. and Karkonen, P., 1969. Studies on the volatile flavour substances in some clover species. Suom. Kemistil., B, 42: 448—451.
Hopkins, P.S. and Pratt, M.S., 1976. Some practical considerations for improving the pregnancy rate of tropical Merinos. Proc. Aust. Soc. Anim. Prod., 11: 153—156.
Hornocker, M.C., 1969. Defensive behaviour in female bighorn sheep. J. Mammal., 50: 128.
Hughes, G.P. and Reid, D., 1952. Studies on the behaviour of cattle and sheep in relation to the utilization of grass. J. Agric. Sci. Camb., 41: 350—366.
Hughes, R.E., Dale, J., Mountford, M.D. and Ellis-Williams, I., 1975. Studies in sheep populations and environment in the mountains of north-west Wales. II. Contemporary distribution of sheep populations and environment. J. Appl. Ecol., 12: 165—178.
Hulet, C.V., Blackwell, R.L. and Ercanbrack, S.K., 1964. Observations on sexually inhibited rams. J. Anim. Sci., 23: 1095—1097.
Hulet, E.V., Ercanbrack, S.K., Price, D.A. and Wilson, L.O., 1962. Mating behaviour of the ewe. J. Anim. Sci., 21: 870—874.
Hunter, R.F., 1954. Some notes on the behaviour of hill sheep. Br. J. Anim. Behav., 2: 75—78.
Hunter, R.F., 1960. Aims and methods in grazing behaviour studies on hill pastures. Proc. 8th Int. Grassl. Congr., pp. 454—458.
Hunter, R.F., 1962. Hill sheep and their pasture: a study of sheep grazing in south-east Scotland. J. Ecol., 50: 651—680.
Hunter, R.F., 1964. Home range behaviour in hill sheep. In: D.J. Crisp (Editor), Grazing in Terrestrial and Marine Environments. Blackwell, Oxford, pp. 155—171.
Hunter, R.F. and Davies, G.E., 1963. The effect of method of rearing on the social behaviour of Scottish Blackface hoggets. Anim. Behav., 5: 183—194.
Hunter, R.F. and Milner, C., 1963. The behaviour of individual, related and groups of south country Cheviot Hill sheep. Anim. Behav., 11: 507—513.
Hutchinson, H.G., Woof, R., Mabon, R.M., Salehe, I. and Robb, J.M., 1962. A study of the habits of Zebu cattle in Tanganyika. J. Agric. Sci. Camb., 59: 301—317.
Hutchinson, J.S.M., O'Connor, P.J. and Robertson, H.A., 1964. Observations on the onset of the breeding season and on the oestrous cycle of the Welsh Mountain ewe. J. Agric. Sci. Camb., 63: 59—60.
Hutchinson, K.J. and McRae, B.H., 1969. Some factors associated with the behaviour and survival of newly shorn sheep. Aust. J. Agric. Res., 20: 512—521.
Imanishi, K., 1950. Social life of semi-wild horses in Tainsaki. III. Summary. Physiol. Ecol., Japan, 4: 29—42.
Inkster, I.J., 1957. The mating behaviour of sheep. Sheep Farming Annu., 1957, pp. 163—169.
Ittner, N.R., Bond, T.E. and Kelly, C.F., 1954. Increasing summer gains of livestock. J. Anim. Sci., 13: 867—877.
Jardine, J.T., 1910. The pasturage system for handling range sheep. U.S. For. Serv. Circular 178, pp. 1—40.
Jardine, J.T. and Anderson, M., 1919. Range Management on the National Forests. U.S. Dep. Agric. Bull. No. 790, 98 pp.
Jones, D.A. and Paddock, R., 1966. Management Notes. You can't turn 'em loose — or can you? J. Range Manage., 19: 96—98.
Jones, E.C. and Barnes, R.J., 1967. Non-volatile organic acids of grasses. J. Sci. Food Agric., 18: 321—323.
Kare, M.R. and Ficken, M.S., 1963. Compariative studies in the sense of taste. In: Y. Zotterman (Editor), Olfaction and Taste. Pergamon Press, Oxford, pp. 285—297.
Kelly, R.W., Allison, A.J. and Shackell, G.H., 1975. Libido testing and subsequent mating performance in rams. Proc. N.Z. Soc. Anim. Prod., 35: 204—212.
Kennedy, G.S. and Holgate, M.D., 1971. The acceptability of phalaris to sheep. Aust. J. Exp. Agric. Husb., 11: 169—172.
Keogh, R.G., 1973a. Herbage growth, and the grazing pattern of sheep set-stocked, on a ryegrass-dominant staggers-prone pasture during summer. N.Z. J. Exp. Agric., 1: 51—54.
Keogh, R.G., 1973b. *Pithanyeus chartarum* spore distribution and sheep grazing patterns in relation to urine-patch and inter-excreta sites within ryegrasss-dominant pastures. N.Z. J. Agric. Res., 16: 353—355.
Kilgour, R., 1972. Some observations on the suckling activity of calves on nurse cows. Proc. N.Z. Soc. Anim. Prod., 32: 132—136.

Kilgour, R. and Campin, D.N., 1973. The behaviour of entire bulls of different ages at pasture. Proc. N.Z. Soc. Anim. Prod., 33: 125—138.
Kilgour, R. and Scott, T.H., 1959. Leadership in a herd of dairy cows. Proc. N.Z. Soc. Anim. Prod., 19: 36—43.
Klopfer, P.H. and Gamble, J., 1966. Maternal "imprinting" in goats: the role of chemical senses. Z. Tierpsychol., 23: 588—592.
Klopfer, P.H. and Klopfer, M.S., 1968. Maternal "imprinting" in goats: fostering of alien young. Z. Tierpsychol., 25: 862—866.
Klopfer, P.H., Adams, K. and Klopfer, M.S., 1964. Maternal "imprinting" in goats. Proc. N.Y. Acad. Sci. U.S.A., 52: 911—914.
Knight, T.W. and Lindsay, D.R., 1973. Identifying the mating performance of individual rams in field flocks. Aust. J. Agric. Res., 24: 579—585.
Kothmann, M.M., 1966. Nutrient content of forage ingested in the morning compared to evening. J. Range Manage., 19: 95—96.
Kropp, J.R., Holloway, J.W., Stephens, D.F., Knori, L., Morrison, R.D. and Totusek, R., 1973. Range behaviour of Hereford, Hereford X Holstein and Holstein non-lactating heifers. J. Anim. Sci., 36: 797—802.
Krueger, W.C., Laycock, W.A. and Price, D.A., 1974. Relationships of taste, smell, sight, and touch to forage selection. J. Range Manage., 27: 258—262.
Lambourne, L.J., 1956. Mating behaviour. Proc. Ruakura Fmrs Conf. Week, Hamilton, N.Z. 1956, pp. 16—20.
Lamond, D.R., 1963. Effect of nutritive status on oestrus, ovulation and Graafian follicles in Merino ewes. Aust. J. Agric. Res., 14: 483—491.
Lampkin, G.H., Quarterman, J. and Kidner, M., 1958. Observations on the grazing habits of grada and Zebu steers in a high altitude temperature climate. J. Agric. Sci. Camb., 50: 211—218.
Lange, R.T., 1969. The piosphere: sheep track and dung patterns. J. Range Manage., 22: 396—400.
Langlands, J.P., 1965. Diurnal variation in the diet selected by free-grazing sheep. Nature (London), 207: 666—667.
Langlands, J.P., 1967. Studies on the nutritive value of the diet selected by grazing sheep. II. Some sources of error when sampling oesophagically fistulated sheep at pasture. Anim. Prod., 9: 167—175.
Langlands, J.P., 1969. Studies on the nutritive value of the diet selected by grazing sheep. IV. Variation in the diet selected by sheep differing in age, breed, sex, strain and previous history. Anim. Prod., 11: 369—378.
Large, R.V. and Spedding, C.R.W., 1976. The agricultural use of amenity grassland. In: S.E. Wright (Editor), Amenity Grasslands: Proceedings of a Conference held at Wye College, Kent, England, March 1976, 45 pp.
Larkin, R.M., 1954. Observations on the grazing behaviour of beef cattle in tropical Queensland. Queensl. J. Agric. Sci., 11: 115—141.
Lees, J.L. and Weatherhead, M., 1970. A note on mating preferences of Clun Forest ewes. Anim. Prod., 12: 173—175.
Leigh, J.H. and Mulham, W.E., 1966a. Selection of diet by sheep grazing semi-arid pastures on the Riverine Plain. I. A bladder saltbush (*Atriplex vesicaria*)—cotton bush (*Kochia aphylla*) community. Aust. J. Exp. Agric. Anim. Husb., 6: 460—467.
Leigh, J.H. and Mulham, W.E., 1966b. Selection of diet by sheep grazing semi-arid pastures on the Riverine Plain. II. A cotton bush (*Kochia aphylla*)—grassland (*Stipa variabilis-Danthonia caespitosa*) community. Aust. J. Exp. Agric. Anim. Husb., 6: 468—474.
Leigh, J.H., Wilson, A.D. and Mulham, W.E., 1968. A study of Merino sheep grazing a cotton bush (*Kochia aphylla*)—grassland (*Stipa variabilis-Danthonia caespitosa*) community on the Riverine Plain. Aust. J. Agric. Res., 19: 947—961.
Le Neindre, P. and Garel, J.P., 1976. Existence d'une période sensible pour l'establissement du compartement matérial de la vache après la mire bais. Biol. Behav., 1: 217—221.
Lent, P.C., 1974. Mother and infant relationships in ungulates. In: V. Geister and F. Walther (Editors), The Behaviour of Ungulates and its Relation to Management. IUCN Publ. New Series No. 24, Morges, Switzerland, pp. 14—55.
Leyhausen, P., 1971. Dominance and territoriality as complemented in mammalian's social structure. In: A.H. Esser (Editor), Behaviour and Environment. The Use of Space by Animals and Man. Plenum Press, New York, N.Y., 400 pp.
Lightfoot, R.J. and Smith, J.A.C., 1968. Studies on the number of ewes joined per ram for flock matings under paddock conditions. I. Mating behaviour and fertility. Aust. J. Agric. Res., 19: 1029—1042.
Lindsay, D.R., 1965. The importance of olfactory stimuli in the mating behaviour of the ram. Anim. Behav., 13: 75—78.
Lindsay, D.R., 1966. Modification of behavioural oestrus in the ewe by social and hormonal factors. Anim. Behav., 14: 73—83.

Lindsay, D.R. and Ellsmore, J., 1968. The effect of breed, season and competition on mating behaviour of rams. Aust. J. Exp. Agric. Anim. Husb., 8: 649—651.

Lindsay, D.R. and Robinson, T.J., 1961a. Studies on the efficiency of mating in the sheep. I. The effect of paddock size and number of rams. J. Agric. Sci. Camb., 57: 137—140.

Lindsay, D.R. and Robinson, T.J., 1961b. Studies on the efficiency of mating in the sheep. II. The effect of freedom of rams, paddock size, and ages of ewes. J. Agric. Sci. Camb., 57: 141—145.

Lofgreen, G.P., Meyer, J.H. and Hull, J.L., 1957. Behaviour patterns of sheep and cattle being fed pasture on sortage. J. Anim. Sci., 16: 773—780.

Louw, D.J., Havenga, C.M. and Hamersma, J., 1948. The walking habits of sheep. Farming S. Afr., 23: 753—755.

Louw, G.N., Steenkamp, C.W.P. and Steenkamp, E.L., 1967. Diet verwantskap tussen die eterekstrakinhoud van karoobossies en hul smaaklikheid vir skape. S. Afr. Tydsk. Landbouwet., 10: 867—873.

Low, W.A., 1972. Community preference by free ranging shorthorns in the Alice Springs district. Proc. Aust. Soc. Anim. Prod., 9: 381—386.

Low, W.A., Hodder, R.M. and Abel, D.E., 1978a. Watering behaviour of British breed cattle in Central Australia. In: K.M.W. Howes (Editor), Studies of the Australian Arid Zone, Part 3 — Water in Rangelands. CSIRO, Melbourne, pp. 165—177.

Low, W.A., Tweedie, R.L., Edwards, C.B.H., Hodder, R.M., Malafant, K.W.J. and Cunningham, R.B., 1978b. The influence of environment on daily maintenance behaviour of free ranging shorthorn cows in central Australia. Appl. Anim. Ethol., (submitted).

Lynch, J.J., 1967. Ranging behaviour of sheep in large areas. Proc. Aust. Ecol. Soc., 2: 167—169.

Lynch, J.J., 1974a. Merino sheep: some factors affecting their distribution in very large paddocks. In: V. Geister and F. Walther (Editors), The Behaviour of Ungulates and its Relation to Management. IUCN Publ. New Series No. 24, Morges, Switzerland, pp. 697—707.

Lynch, J.J., 1974b. Behaviour of sheep and cattle in the more arid areas of Australia. In: A.D. Wilson (Editor), Studies of the Australian Arid Zone, Part 2 — Animal Production. CSIRO, Melbourne, pp. 37—49.

Lynch, J.J. and Alexander, G., 1976. The effect of gramineous windbreaks on behaviour and lamb mortality among shorn and unshorn Merino sheep during lambing. Appl. Anim. Ethol., 2: 305—325.

McBride, G., Arnold, G.W., Alexander, G. and Lynch, J.J., 1967. Ecological aspects of the behaviour of domestic animals. Proc. Ecol. Soc. Aust., 2: 133—165.

McCann, L.J., 1956. Ecology of the mountain sheep. Am. Midl. Nat., 56: 297—324.

McClymont, G.L., 1967. Selectivity and intake in the grazing ruminant. In: C.F. Cole (Editor), Handbook of Physiology, Section 6. Alimentary Canal 1. Am. Physiol. Soc., Washington, D.C., pp. 129—137.

McDaniel, A.H. and Roark, C.B., 1956. Performance and grazing habits of Hereford and Aberdeen-Angus cows and calves on improved pastures as related to types of shade. J. Anim. Sci., 11: 59—63.

McDougall, P., 1975. The feral goats of Kielderhead Moor. J. Zool. Lond., 176: 215—246.

McHugh, J.F. and Edwards, M.S.H., 1958. Lamb loss investigations — Rutherglen Research Station. J. Agric. (Vic., Aust.), 56: 425—438.

McIlvain, E.H. and Shoop, M.C., 1965a. Gain response of steers to change of pasture and social order. Progress in Range Research in 17 Western States. H.J. Hodgson (Compiler). Cooperative State Research Service, USDA, Washington, D.C., p. 246.

McIlvain, E.H. and Shoop, M.C., 1965b. Methods for distributing livestock grazing pressure. Progress in Range Research in 17 Western States. H.J. Hodgson (Compiler). Cooperative State Research Service, USDA, Washington, D.C., p. 247.

McIlvain, E.H. and Shoop, M.C., 1971. Shade for improving cattle gains and rangeland use. J. Range Manage., 24: 181—184.

MacLusky, D.S., 1959. Drinking habits of grazing cows. Agriculture Lond., 66: 383—386.

Malechek, John C. and Smith, B.M., 1976. Behaviour of range cows in response to winter weather. J. Range Manage., 29: 9—12.

Mann, T.L.J., 1968. A comparison of lamb survival in fox proof and unprotected enclosures. Proc. Aust. Soc. Anim. Prod., 7: 250—254.

Marincowitz, G., Pretorius, P.S. and Herbst, S.N., 1966. Dominance in sexual behaviour and mating dexterity in Merino rams. S. Afr. J. Agric. Sci., 9: 971—980.

Marsh, R. and Campling, R.C., 1970. Fouling of pastures by dung. Herb. Abstr., 40: 123—130.

Marten, G.C. and Donker, J.D., 1964. Selective grazing induced by animal excreta. I. Evidence of occurrence and superficial remedy. J. Dairy Sci., 47: 773—776.

Marten, G.C., Jordan, R.M. and Hovin, A.W., 1976. Biological significance of reed canarygrass alkaloids and associated palatability variation to grazing sheep and cattle. Agron. J., 68: 909—914.

Mattner, P.E., Braden, A.W.H. and Turnbull, K.W., 1967. Studies in flock mating of sheep. I. Mating behaviour. Aust. J. Exp. Agric. Anim. Husb., 7: 103—109.

Mattner, P.E., Braden, A.W.H. and George, J.M., 1971. Studies in flock mating of sheep. IV. The relation of libido tests to subsequent service activity of young rams. Aust. J. Exp. Agric. Anim. Husb., 11: 473—477.

Mattner, P.E., Braden, A.W.H. and George, J.M., 1973. Studies in flock mating of sheep. V. Incidence, duration and effect on fertility of mutual sexual inactivity in young rams. Aust. J. Exp. Agric. Anim. Husb., 13: 35—40.
Merrill, L.B. and Taylor, C.A., 1976. Take note of the versatile goat. Rangeman's J., 3: 74—76.
Miller, G.R., 1968. Some responses of hill ewes and lambs to artificial shelter. Anim. Prod., 10: 59—66.
Milton, W.E.J., 1953. The palatability of herbage on undeveloped grasslands in west-central Wales. Emp. J. Exp. Agric., 21: 116—122.
Moore, R.W., 1966. "Ewe X side of udder" interactions in milk yield in relation to the number of lambs suckled. Nature (London), 211: 663—664.
Moorefield, J.G. and Hopkins, H.H., 1951. Grazing habits of cattle in a mixed-prairie pasture. J. Range Manage., 4: 151—157.
Moran, A.B., 1974. Mating pattern of maiden crossbred ewes mated alone or in flocks of mixed ages. Proc. Aust. Soc. Anim. Prod., 10: 279—282.
Morgan, P.D., 1970. Maternal Behaviour in the Ewe. Ph.D. Thesis, Univ. Western Australia, 209 pp.
Morgan, P.D. and Arnold, G.W., 1972. Changes in walking activity of the ewe before and after parturition. J. Aust. Inst. Agric. Sci., 38: 140.
Morgan, P.D. and Arnold, G.W., 1974. Behavioural relationships between Merino ewes and lambs during the four weeks after birth. Anim. Prod., 19: 196—206.
Morgan, P.D., Arnold, G.W. and Lindsay, D.R., 1972a. A note on the mating behaviour of ewes with various senses impaired. J. Reprod. Fertil., 30: 151—152.
Morgan, P.D., Arnold, G.W. and Boundy, C.A.P., 1972b. The behaviour of ewes and their new born lambs in a hot climate. Proc. Aust. Soc. Anim. Prod., 9: 371—375.
Morgan, P.D., Boundy, C.A.P., Arnold, G.W. and Lindsay, D.R., 1975. The roles played by the senses of the ewe in the location and recognition of lambs. Appl. Anim. Ethol., 1: 143—150.
Moseley, T.M., 1929. Dairy work at the Huntley Field Stn, Huntley, Mont. 1918—1927. USDA Tech. Bull. 116, 48 pp.
Moule, G.R., 1950. The influence of rapid decrease in the hours of daylight on the sexual desire of Merino rams. Aust. Vet. J., 25: 84—87.
Moule, G.R. and Young, R.B., 1961. Field observations on the daily milk intake of Merino lambs in semi-arid tropical Queensland. Queensl. J. Agric. Sci., 18: 221—229.
Mueggler, W.F., 1965. Cattle distribution on steep slopes. J. Range Manage., 18: 255—257.
Müller, W.J., Low, W.A., Lendon, C. and Dudzinski, M.L., 1976. Variation in grazing patterns of free ranging cattle in a semi-arid area. Proc. Aust. Soc. Anim. Prod., 11: 461—464.
Munro, J., 1956. Observations on the suckling behaviour of young lambs. Br. J. Anim. Behav., 4: 34—36.
Munro, J., 1961. Shelter for livestock: A survey of two hill areas in Scotland. Hill Farming Res. Org., Penicuik, 2nd Rep., 1961, pp. 83—91.
Munro, J., 1962. The use of natural shelter by hill sheep. Anim. Proc., 4: 343—349.
Musangi, R.S., 1965. Feed intake studies in ruminants. II. The grazing behaviour of Friesian and Nganda steers on tropical pasture. Afr. Soils, 10: 321—324.
Musangi, R.S., 1966. The influence of supplementary feeding and grazing restriction on the voluntary intake of roughage and liveweight gain in beef cattle. E. Afr. Agric. For. J., 31: 272—275.
Nicholls, P.J., Fowler, D.G. and Kearings, R.D., 1976. The effects of paddock size, liveweight change and flock dispersion on fertility of Merino ewes. Proc. Aust. Soc. Anim. Prod., 11: 29.
Nicol, A.M. and Sharafeldin, M.A., 1975. Observations on the behaviour of single suckled calves from birth to 120 days. Proc. N.Z. Soc. Anim. Prod., 35: 221—230.
Obioha, F.C., Clanton, D.C., Rittenhouse, L.R. and Streeter, C.L., 1970. Sources of variation in chemical composition of forage ingested by esophageal fistulated cattle. J. Range Manage., 23: 133—136.
Odberg, F.O., 1975. Eliminative and grazing behaviour in the horse. Appl. Anim. Ethol., 1: 208 (Abstr.).
Ödberg, F.D. and Francis-Smith, K., 1976. A study on eliminative and grazing behaviour — the use of the field by captive horses. Equine Vet. J., 8: 147—149.
Ödberg, F.O. and Francis-Smith, K., 1977. Studies on the formation of ungrazed eluminative areas in fields used by horses. Appl. Anim. Ethol., 3: 27—34.
O'Donovan, P.D., Barnes, R.F., Plumlee, M.P., Packett, L.V. and Mott, G.O., 1967. Soluble and structural components of alfalfa (*Medicago sativa* L.) and selected Reed Canary grass (*Phalaius arundunacea* L.) clones. Agron. J., 59: 478—481.
Oh, H.K., Sakai, T., Jones, M.B. and Longhurst, W.M., 1967. Effect of various essential oils updated from douglas fir needles upon sheep and deer rumen in microbial activity. Appl. Microbiol., 15: 777—784.
Ozanne, P.G. and Howes, K.M.W., 1971. Preference of grazing sheep for pasture of high phosphate content. Aust. J. Agric. Res., 22: 941—950.
Payne, W.J.A., Laing, W.I. and Raikova, E.N., 1951. Grazing behaviour of dairy cattle in the tropics. Nature (London), 167: 610—611.
Peacock, F.M., Kirk, W.G., Hodges, E.M., Palmer, A.Z. and Carpenter, J.W., 1965. Influence of sum-

mer pasture, diethylstilbestrol, and shade on fattening cattle in South Florida. Tech. Bull. 700. Agric. Exp. Stns, Inst. of Food and Agric. Sci., Univ. of Florida, Gainesville, USA, pp. 1—14.
Pepelko, W.E. and Clegg, M.T., 1965. Influence of season of the year upon patterns of sexual behaviour in male sheep. J. Anim. Sci., 24: 633—637.
Perkins, D.F., Crook, I.G. and Thomson, A.G., 1970. A census of domestic livestock grazing on Dartmoore. Natural Environ. Res. Council. Report on the Dartmoor Ecological Survey 1969. Nature Conservancy, Bangor, pp. 41—45.
Petersen, B., 1956. The behaviour of identical bovine twins. Zool. Anz. Suppl., 20: 91—95.
Petersen, R.G., Lucas, H.L. and Woodhouse, W.W., 1956. The distribution of excreta by freely grazing cattle and its effect on pasture fertility. I. Excreta distribution. Agron. J., 48: 449—448.
Peterson, R.A. and Woolfolk, E.J., 1955. Behaviour of Hereford cows and calves on short grass range. J. Range Manage., 8: 51—57.
Plice, M.J., 1952. Sugar versus the intuitive choice of foods by livestock. J. Range Manage., 5: 69—74.
Poindron, P., 1974. Etude de la relation mere-jeune chez des brebis (*Ovis aries*), lars de l'allaitement. C.R. Acad. Sci. Paris t. 278, Seria D: 2691—2694.
Poindron, P. and Carrick, M.J., 1976. Hearing recognition of the lamb by its mother. Anim. Behav., 24: 600—602.
Powell, T.L., 1966. Components of climate in relation to walking and grazing activity of sheep. Proc. 9th Int. Cong. Anim. Prod., Edinburgh, Oliver and Boyd, 120 pp.
Powell, T.L., 1967. A note on relationships between components of weather on the location of defaecation in outwintered cattle. Anim. Prod., 9: 413—416.
Quinn, J.A. and Harvey, D.F., 1970. Trampling losses and travel by cattle on sandhills range. J. Range Manage., 23: 50—55.
Rankine, G. and Donaldson, L.E., 1968. Animal behaviour and calf mortalities in a North Queensland breeding herd. Proc. Aust. Soc. Anim. Prod., 7: 138—143.
Rawes, M. and Welch, D., 1964. Studies on sheep grazing in the northern Pennines. J. Br. Grassl. Soc., 19: 403—411.
Rawes, M. and Welch, D., 1966. Further studies on sheep grazing in the northern Pennines. J. Br. Grassl. Soc., 21: 56—61.
Rawes, M. and Welch, D., 1969. Upland productivity of vegetation and sheep at Moor House National Nature Reserve, Westmoorland, England. Oikos Suppl., II: 14—72.
Reid, J.L. and Jung, G.A., 1965. Influence of fertilizer treatment on intake, digestibility and palatability of fall fescue hay. J. Anim. Sci., 25: 615—625.
Reid, R.L., Jung, G.A. and Murray, S.J., 1966. Nitrogen fertilization in relation to palatability and nutritive value of orchard grass. J. Anim. Sci., 25: 636—645.
Reid, R.L., Jung, G.A. and Kinsey, C.M., 1967. Nutritive value of nitrogen fertilized orchard grass pasture at different periods of the year. Agron. J., 59: 519—525.
Reppert, J.N., 1960. Forage preference and grazing habits of cattle at the Eastern Colorado Range Station. J. Range Manage., 13: 58—64.
Rheingold, H.L., 1963. Introduction. In: H.L. Rheingold (Editor), Maternal Behaviour in Mammals. John Wiley & Sons Inc., New York, N.Y., 349 pp.
Rhoad, A.O., 1938. Some observations on the response of purebred *Bos taurus* and *Bos indicus* cattle and their crossbred types to certain conditions of the environment. Proc. Am. Soc. Anim. Prod., 38: 284—295.
Riches, J.A. and Watson, R.H., 1954. The influence of the introduction of rams on the incidence of oestrus in Merino ewes. Aust. J. Agric. Res., 5: 141—155.
Ricordeau, G., Boccard, R. and Denamur, R., 1960. Mesure de la production laitière des brebis pendant la période d'allaitement. Ann. Zootech., 9: 97—120.
Riney, T. and Caughley, G., 1959. A study of home range in a feral goat herd. N.Z. J. Sci., 2: 157—170.
Robertson, H.A. and Rakha, A.M., 1965. Time of onset of oestrus in the ewe. J. Reprod. Fertil., 10: 271—272.
Roe, R. and Mottershead, B.E., 1962. Palatability of *Phalaris arundinacea.* Nature (London), 193: 255—257.
Rollinson, D.H.L., 1960. Reproductive habits and fertility of indigenous cattle to artificial insemination in Uganda. J. Agric. Sci. Camb., 60: 279—284.
Rossdale, P.D., 1968. Perinatal behavior in the thoroughbred horse. In: M.W. Fox (Editor), Abnormal Behavior in Animals. Washington University, St. Louis, Mo., 563 pp.
Rutter, N., 1968. Time lapse photographic studies of livestock behaviour outdoors on the College Farm Aberystwyth. J. Agric. Sci. Camb., 71: 257—265.
Schaller, G.B. and Laurie, A., 1974. Courtship behaviour in the wild goat. Sonderdruck aus Z. Saeugetierkd., 39 (2): 125—127.
Schein, M.W. and Fohrman, M.H., 1955. Social dominance relationships in a herd of dairy cattle. Br. J. Anim. Behav., 3: 45—55.
Schmidt, P.J., 1969. Observations on the behaviour of cattle in a hot dry region of the Northern Territory of Australia, with particular reference to walking, watering and grazing. M.Sc. Thesis, The

University of New England, Armidale, N.S.W., 220 pp.
Schoen, A.M.S., Banks, E.M. and Curtis, S.E., 1976. Behaviour of young Shetland und Welsh ponies (*Equus caballus*). Biol. Behav., 1: 192—216.
Schloeth, V.R., 1961. Das Sozialleben des Camargue-Rindes. Qualitative und quantitative Untersuchungen über die sozialen Beziehungen — insbesondere die soziale Rangordnung — des halbwilden französischen Kampfrindes. Z. Tierpsychol., 18: 547—627.
Scott, J.P., 1969. Introduction to Animal Behaviour. In: E.S.E. Hafez (Editor), Behaviour of Domestic Animals. Ballière, Tindall and Cassell, London, Great Britain, 647 pp.
Seath, D.M. and Miller, G.D., 1946. Effect of warm weather on grazing performance of milking cows. J. Dairy Sci., 29: 199—205.
Seath, D.M. and Miller, G.D., 1947. Effect of hay feeding in summer on milk production and grazing performance of dairy cows. J. Dairy Sci., 39: 921—925.
Selman, I.E., McEwan, A.D. and Fisher, E.W., 1970a. Studies on natural suckling in cattle during the first eight hours post partum. I. Behavioural studies (dams). Anim. Behav., 18: 276—283.
Selman, I.E., McEwan, A.D. and Fisher, E.W., 1970b. Studies on natural suckling in cattle during the first eight hours post partum. II. Behavioural studies (calves). Anim. Behav., 18: 284—289.
Shackell, G.H., Kelly, R.W. and Allison, A.J., 1977. Seasonal variation in libido of rams. N.Z. J. Exp. Agric., 5: 721—722.
Sharafeldin, M.A. and Shafie, M.M., 1965. Animal behaviour in the sub-tropics. II. Grazing behaviour of sheep. Neth. J. Agric. Sci., 13: 239—247.
Shelley, L., 1970. Interrelationships between the duration of parturition post-natal behaviour of ewes and lambs, and the incidence of neonatal mortality. Proc. Aust. Soc. Anim. Prod., 8: 348—352.
Shepperd, A.J., Blaser, R.E. and Kincaid, C.M., 1957. The grazing habits of beef cattle on pasture. J. Anim. Sci., 16: 681—687.
Shepperd, J.H., 1921. The trail of the short grass steer. N. D. Agric. Coll. Bull. 154, p. 8.
Shillito, E.E., 1975. A comparison of the role of vision and hearing in lambs finding their own dams. Appl. Anim. Ethol., 1: 369—377.
Shillito, E.E. and Alexander, G., 1975. Mutual recognition amongst ewes and lambs of four breeds of sheep (*Ovis aries*). Appl. Anim. Ethol., 1: 151—165.
Shillito, E.E. and Hoyland, V.J., 1971. Observations on parturition and maternal care in Soay sheep. J. Zool. Lond., 165: 509—512.
Shreffler, C. and Hohenboken, W.D., 1974. Dominance and mating behaviour in ram lambs. J. Anim. Sci., 39: 725—731.
Simons, A.B. and Marten, G.C., 1971. Relations of indole alkaloids to palatability of *Phalaris arundinacea* L. Agron. J., 63: 915—919.
Smith, C.A., 1961. Studies on the Northern Rhodesia Hyparrhenia veld. III. The effect on the growth and grazing-behaviour of indigenous cattle on restricting their daily grazing time by night kraaling. J. Agric. Sci. Camb., 56: 243—248.
Smith, D.R. and Lang, R.L., 1958. The effect of nitrogenous fertilizers on cattle distribution on mountain range. J. Range Manage., 11: 218—249.
Smith, F.V., 1965. Instinct and learning in the attachment of lamb and ewe. Anim. Behav., 13: 84—86.
Smith, F.V., Van-Toller, C. and Boyes, T., 1966. The 'critical period' in the attachment of lambs and ewes. Anim. Behav., 14: 120—125.
Smoliak, S. and Peters, H.F., 1955. Climatic effects on foraging performance of beef cows on winter range. Can. J. Agric. Sci., 35: 213—216.
Sneva, F.A., 1970. Behavior of yearling cattle on eastern Oregon range. J. Range Manage., 23: 155—158.
Sojetado, R.M., 1952. The growth and habits of lambs. Philipp. Agric., 35: 572—578.
Southcott, W.H., Hill, M.K., Watkin, B.R. and Wheeler, J.L., 1968. Effect of paddock size, stocking rate, anthelmintics, and trace elements on the weight gain of young cattle. Proc. Aust. Soc. Anim. Prod., 7: 118—122.
Spencer, C.C., 1943. Notes on the life history of Rocky Mountain Bighorn sheep in the Tarryall Mountains of Colorado. J. Mammal., 24: 1—10.
Squires, V.R., 1971. Temporal patterns of activity in a small flock of Merino sheep as determined by an automatic recording technique. Anim. Behav., 19: 657—660.
Squires, V.R., 1974. Grazing distribution and activity patterns of Merino sheep on a saltbush community in south-east Australia. Appl. Anim. Ethol., 1: 17—30.
Squires, V.R., 1975a. Ecology and behaviour of domestic sheep (*Ovis aries*): a review. Mammal Rev., 5: 35—57.
Squires, V.R., 1975b. Environmental heterogeneity as a factor in group size determination among grazing sheep. Proc. N.Z. Soc. Anim. Prod., 35: 184—190.
Squires, V.R., 1976. Walking, watering and grazing behaviour of Merino sheep on two semi-arid rangelands in south-west New South Wales. Aust. Rangel. J., 1: 13—23.
Squires, V.R. and Wilson, A.D., 1970. Distance between food and water supply and its effect on drinking frequency and food and water intake of Merino and Border Leicester X Merino sheep. Aust. J. Agric. Res., 22: 283—290.

Squires, V.R., Wilson, A.D. and Daws, G.T., 1972. Comparison of walking behaviour of some Australian sheep. Proc. Aust. Soc. Anim. Prod., 9: 376—380.

Stapleton, R.G., 1934. Palatability and the management of the poorer grasslands. J. Minist. Agric. (G.B.), 41: 321—328.

Staten, H.W., 1949. Palatability trials of winter pasture crops and effect of phosphate fertilizers on palatability. Okla. Agric. Exp. Stn, Tech. Bull. T—35, 10 pp.

Stewart, J., 1953. Grazing habits and trace elements. Br. J. Anim. Behav., 1: 116—117.

Stobbs, T.H., 1973a. The effect of plant structure on the intake of tropical pastures. I. Variation in the bite size of grazing cattle. Aust. J. Agric. Res., 24: 809—819.

Stobbs, T.H., 1973b. The effect of plant structures on the intake of tropical pastures. II. Differences in sward structure, nutritive value, and bite size of animals grazing *Setaria anceps* and *Chloris gayana* at various stages. Aust. J. Agric. Res., 24: 821—829.

Stobbs, T.H., 1974. Components of grazing behaviour of dairy cows on some tropical and temperate pastures. Proc. Aust. Soc. Anim. Prod., 10: 299—302.

Strassia, C.A., Thorn, M., Rice, R.W. and Smith, D.R., 1970. Grazing habits, diet and performance of sheep on alpine ranges. J. Range Manage., 23: 200—208.

Sumner, S.L., Ancalmo, R. and Warnick, A.C., 1968. Behaviour of bulls and rams during the breeding season. J. Anim. Sci., 27: 1197.

Suzuki, S., Shinde, Y., Yoshii, K. and Takimoto, Y., 1967. Studies on the eliminative behaviour of dairy cows. Res. Bull. Obihiro Zootech. Univ. Ser. 1, 5: 45—54.

Tayler, J.C., 1953. The grazing behaviour of bullocks under two methods of management. Br. J. Anim. Behav., 1: 72—77.

Taylor, E.L., 1954. Grazing behaviour and helminthic disease. Br. J. Anim. Behav., 2: 61—62.

Taylor, R.H., Bell, B.D. and Wilson, P.R., 1970. Royal albatrosses, feral sheep and cattle on Campbell Island. N.Z. J. Sci., 13: 78—88.

Thiele-Wittig, H.C., 1974. Maintenance of previously cultivated land not now used for agriculture. Agric. Environ., 1: 129—137.

Tomkins, T. and Bryant, M.J., 1972. Mating behaviour in a small flock of lowland sheep. Anim. Prod., 5: 203—210.

Tribe, D.E., 1949. Some seasonal observations on the grazing habits of sheep. Emp. J. Exp. Agric., 17: 105—115.

Tribe, D.E., 1950. Influence of pregnancy and social facilitation on the behaviour of the grazing sheep. Nature (London), 166: 74.

Tschanz, B., 1962. The relation between mother animal and young in the Mouflon (*Ovis aries musimon*, Pall.). Experientia, 18(4): 187—190.

Tyler, S.J., 1972. The behaviour and social organization of the New Forest ponies. Anim. Behav. Mon., Part 5: 87—196.

Valentine, K.A., 1947. Distance from water as a factor in grazing capacity of rangeland. J. For., 45: 749—754.

Van Dyne, G.M. and Heady, H.F., 1965a. Dietary chemical composition of cattle and sheep grazing annual range. J. Range Manage., 18: 78—85.

Van Dyne, G.M. and Heady, H.F., 1965b. Botanical composition of sheep and cattle diets on a mature annual range. Hilgardia, 36: 465—492.

Van Dyne, G.M. and Van Horn, J.L., 1965. Distance travelled by sheep on winter range. Proc., West. Sect., Am. Soc. Anim. Sci., 16: 1—6.

Von Sambraus, H.H., 1968. Sexual behaviour of free running cattle. Proc. 6th Congr. Int. Reprod. Insem. Artif., 2: 1667—1670.

Von Sambraus, H.H., 1977. Observations and experiments on social behaviour in a herd of cattle during an 11-year period. Appl. Anim. Ethol., 3: 199—200 (Abstr.).

Von Sambraus, H.H. and Waring, G.H., 1975. Der Einfluss des Harns brünstiger Kühe auf die geschlechtslust von Stieren. Sonderdruck aus Z. Saeugetier Kol., 40: 49—54.

Wagnon, K.A., 1963. Behavior of beef cows on a Californian range. Calif. Agric. Exp. Stn Bull. 799, 58 pp.

Wagnon, K.A., 1965. Social dominance in range cows and its effect on supplemental feeding. Calif. Agric. Exp. Stn Bull. 819, 31 pp.

Wagnon, K.A., Guilbert, H.R. and Hart, G.H., 1959. Beef cattle investigations on the San Joaquin Experimental Range. Calif. Agric. Exp. Stn Bull. 765, 71 pp.

Wagnon, K.A., Loy, R.G., Rollins, W.C. and Carroll, F.D., 1966. Social dominance in a herd of Angus, Hereford, and Shorthorn cows. Anim. Behav., 14: 474—479.

Waite, R., MacDonald, W.B. and Holmes, W., 1951. Studies in grazing management. III. The behaviour of dairy cows grazed under the close-folding and rotational systems of management. J. Agric. Sci. Camb., 41: 163—173.

Walker, D.E., 1962. Suckling and grazing behaviour of beef heifers and calves. N.Z. J. Agric. Res., 5: 331—338.

Walker, D.M., 1950. Observations on behaviour in young calves. Bull. Anim. Behav., 1: 5—10.

Walkley, J.R.W. and Barber, A.A., 1976. The relationship between libido score and fertility in Merino rams. Proc. Aust. Soc. Anim. Prod., 11: 141—144.
Wallace, J.D., Free, J.C. and Denham, A.H., 1972. Seasonal changes in herbage and cattle diets on sandhill grassland. J. Range Manage., 25: 100—104.
Wallace, L.R., 1949. Observations of lambing behaviour in ewes. N.Z. Soc. Anim. Prod., 9: 85—96.
Watson, R.H., Alexander, G., Cumming, I.A., McDonald, J.W., McLaughlin, J.W., Rizzoli, D.J. and Williams, D., 1968. Reduction of perinatal loss of lambs in winter in western Victoria by lambing in sheltered individual pens. Proc. Aust. Soc. Anim. Prod., 7: 243—249.
Weaver, J.E. and Tomanek, G.W., 1951. Ecological studies in a midwestern range: The vegetation and effects of cattle on its composition and distribution. Nebr. Conserv. Serv. Bull. 31, 82 pp.
Weir, W.C. and Torrell, D.T., 1959. Selective grazing by sheep as shown by a comparison of the chemical composition of range and pasture forage obtained by hand clipping and that collected by oesophageal-fistulated sheep. J. Anim. Sci., 18: 641—649.
Welch, D. and Rawes, M., 1966. The intensity of sheep grazing on high-level blanket bog in Upper Teesdale. Ir. J. Agric. Res., 5: 185—196.
Welch, R.A.S. and Kilgour, R., 1972. A survey of lambing practices in stud sheep flocks in New Zealand. Proc. N.Z. Soc. Anim. Prod., 32: 115—122.
Whateley, J., Kilgour, R. and Dalton, D.C., 1974. Behaviour of hill country sheep breeds during farming routines. Proc. N.Z. Soc. Anim. Prod., 34: 28—36.
Wheeler, J.L., Rearden, T.F. and Lambourne, L.J., 1963. The effect of pasture availability and shearing stress on herbage intake of grazing sheep. Aust. J. Agric. Res., 14: 364—372.
Wilkins, H.L., Bates, R.P. and Henson, P.R., 1953. Tannin and palatability in Sericea Leispedeza (L. *cuneata*). Agron. J., 45: 335—336.
Wilson, A.D., 1974. Water consumption and water turnover of sheep grazing semi-arid pasture communities in New South Wales. Aust. J. Agric. Res., 25: 339—347.
Wilson, A.D., 1976. Comparison of sheep and cattle grazing on a semi-arid grassland. Aust. J. Agric. Res., 27: 155—162.
Wilson, A.D., Leigh, J.H., Hindley, N.L. and Mulham, W.E., 1975. Comparison of the diets of goats and sheep on a *Casuarina cristata-Heterodendrum oleifolium* woodland community in western New South Wales. Aust. J. Exp. Agric. Anim. Husb., 15: 45—53.
Wilson, P.N., 1961. The grazing behaviour and free-water intake of East African shorthorned Zebu heifers at Serere, Uganda. J. Agric. Sci. Camb., 56: 351—364.
Wilson, P.R. and Orwin, D.F.G., 1964. The sheep population of Campbell Island. N.Z. J. Sci., 7: 460—490.
Winfield, C.G., 1970. The effect of stocking intensity at lambing on lamb survival and ewe and lamb behaviour. Proc. Aust. Soc. Anim. Prod., 8: 29—32.
Winfield, C.J., Brown, W. and Lucas, I.A.M., 1969. Sheltering behaviour at lambing by Welsh Mountain ewes. Anim. Prod., 11: 101—105.
Wohanka, K., 1962. Beobachtungen bei der freien Paarung des Hausrindes. Zuchthyg. Fortpflstor. Besamung Haustiere, 6: 315—336.
Woods, D.L. and Clarke, K.W., 1974. Palatability of Reed Canary Grass pasture. Can. J. Pl. Sci., 54: 89—91.
Woolfolk, E.J., 1949. Stocking Northern Great Plains sheep range for sustained high production. USDA Circular 804, April 1949.
Yeates, N.T.M., 1965. The influence of climate. In: N.T.M. Yeates (Editor), Modern Aspects of Animal Production. Butterworths, London, pp. 89—186.

INDEX